Praise for *In the Blink of an*

"A well-written book, containing much really interesting science and a good strong hypothesis that will surely stimulate others to praise, to criticize and try to refine or replace."

–*Washington Post Book World*

" Parker's research has that pop-science 'wow' factor—dramatic transformations over aeons of time, alien-like life forms, fragments of the secret of our own emergence."

–*Village Voice*

"Compelling."

–*Science News*

"[Parker is a] genius, a cool-headed logician with the soul of an artist. . . . [He] has managed to crack a mystery that evolutionists have fretted over since Darwin first sharpened his quill . . . *In the Blink of an Eye* might very well make him a celebrity."

–*Seed*

"Parker's ideas are fascinating."

–*Boston Globe*

"Parker will have more than a few palaeontologists choking on their cornflakes."

–*New Scientist*

"The outlines of [Parker's] argument are laid out with compelling logic and clarity, and his solution to the Cambrian mystery seems both brilliant and obvious: we must have been blind to miss it."

–*London Sunday Telegraph*

"*In the Blink of an Eye* presents its arguments the way a prosecutor presents a criminal case against the accused in a courtroom melodrama. . . . I don't think you can find a more reader-friendly introduction to evolutionary biology."

–*San Jose Mercury News*

"Full of fascinating scientific lore . . . The flash of unexpected insight that characterizes [Parker's] discovery is of the rarest kind, and with a book like *In the Blink of an Eye*, readers have a chance to share in one of those 'aha!' moments that happen so infrequently in the world of science."

–*Readerville Journal*

"[Parker's] clarity will thrill science fans, as will his revolutionary theory."

–*Booklist*

"Parker's conclusion is both convincing and surprisingly fresh . . . Compelling . . . Cutting-edge science, highly recommended."

–*Kirkus* (starred review)

"An informative work of easily accessible science."

–*Boston Herald*

"A young, brash zoologist . . . Parker makes a compelling case."

–*San Diego Union Tribune*

"An insightful glimpse into the mind of the scientist. . . [A] thought-provoking work."

–*Library Journal*

"A brilliant and eminently readable evolutionary detective tale . . . Parker's energy and intelligence are undeniable . . . [He] has led us down a remarkable trail and one hopes he has many others to explore."

–*Roanoke Times*

"[Parker's] central argument certainly deserves careful attention . . . fascinating examples."

–*American Scientist*

In the Blink of an Eye

ANDREW PARKER

A Member of the Perseus Books Group
New York

Published by Basic Books,
A Member of the Perseus Books Group

Hardback first published in 2003 by Perseus Publishing
Paperback first published in 2004 by Basic Books

Books published by Basic Books are available at special discounts for bulk purchases in the United States by corporations, institutions, and other organizations. For more information, please contact the Special Markets Department at the Perseus Books Group, 11 Cambridge Center, Cambridge MA 02142, or call (617) 252–5298, (800) 255–1514 or e-mail special.markets@perseusbooks.com.

Library of Congress Cataloging-in-Publication Data
is available from the Library of Congress
ISBN 0–7382–0607–5 (hc) ISBN 0–465–05438–2 (pbk)

First Edition
1 2 3 4 5 6 7 8 9 10—06 05 04

To my parents

Contents

List of Illustrations

Integrated illustrations

Plate section

When you have eliminated the impossible, whatever remains, however improbable, must be the truth

SIR ARTHUR CONAN DOYLE, *A Study in Scarlet* (1887)

Preface

> The case [for the abrupt appearance of Cambrian fossils] at present must remain inexplicable . . . and may be truly urged as a valid argument against the views [on evolution] here entertained
>
> CHARLES DARWIN, *On the Origin of Species* (sixth and final edition, 1872)

The Big Bang in animal evolution was perhaps the most dramatic event in the history of life on Earth. During this blink of an eye in such history, all the major animal groups found today evolved hard parts and became distinct shapes, simultaneously and for the first time. This happened precisely 543 million years ago, at the beginning of a period in geological history called the Cambrian, and so has become known as the 'Cambrian explosion'. But what lit the Cambrian fuse?

Until now, we have been without an acceptable explanation for this extraordinary burst in evolution – there is strong evidence against all the contending theories put forward. If time is given to consider most previous explanations, it becomes clear that in fact they explain a different evolutionary event and not the Cambrian explosion, as will be introduced early on in this book. That these two events were once amalgamated had been extremely misleading. In short, we know very well *what* happened during evolution's Big Bang, indeed numerous books have already been written on this question, but we don't know *why* it happened. *Why* it happened is the puzzle this book sets out to solve.

The mention of a 'puzzle' and a 'search for clues' is very appropriate to the story behind the discovery of the *why*, and this book grew naturally into a detective story. After all, this topic will emerge as real scientific crime. I have spent many years stumbling into different fields of science, and it was while travelling along this uneven road that I ended up at the doorstep of the Cambrian. Almost by themselves, the clues towards a Cambrian theory just kept on accumulating, and eventually, after there were still no signs of evidence to the contrary, I became satisfied that the 'truth had remained'.

To uncover the real cause of the Cambrian explosion *all* the pieces of the puzzle are needed. After introducing the problem in Chapter 1, the following seven chapters of this book will be dedicated to the more significant pieces. In the course of these chapters a multidimensional picture will be painted showing how life works today, what happened during the course of evolution on Earth and, consequently, how life worked at different times in the geological past. Having been warned that the more technical terms I adopt the smaller my audience will be, I have responded by keeping scientific names and terminology to a minimum. I have tried to use, or even invent, common names of animals wherever possible, and must apologise if this method appears too simplistic or distracting. Nonetheless, the most important, recurring scientific terms have necessarily survived the editorial process.

By the beginning of the penultimate chapter, all the clues needed to solve the *why* of the Cambrian explosion will have been presented. Scientific evidence will be extracted not just from biology, but also geology, physics, chemistry, history and art. Subjects such as eyes, colour, fossils, predators, Egyptian statues, the deep sea and coral reefs will be entertained. What was the significance of Maxwell's breakfast or of Newton's peacock to our understanding of evolution? Might they be on a par with Charles Doolittle Walcott's monumental discovery of the Cambrian 'Burgess Shale' fossils themselves? I feel that the Cambrian explosion is something worthy of anyone's time, and that the explanation of this event is worthwhile publicising. I hope readers will agree.

My road to the Cambrian was possible only because of some wonderful opportunities presented to me, for which I am extremely grateful. In the first place there were Penny Berents and Pat Hutchings, who offered me my first position at the Australian Museum in Sydney. Here I was lucky enough to spend several years examining living and preserved specimens from every major animal group on Earth – an experience which contrasted greatly with my days studying animal diversity from a textbook as an undergraduate. Then there were Jim Lowry and Noel Tait, at the Australian Museum and Macquarie University (Sydney) respectively, who registered my research for a Ph.D. degree, and taught me so much about animal diversity, ecology and evolution. But I also received considerable help and encouragement from many more members of the Australian Museum than I have space to list here. I am grateful to them all.

By now I had chosen to study seed-shrimps as my specialist subject, and received expert tuition from Lou Kornicker at the Smithsonian Institution (Washington, DC) and Anne Cohen (Los Angeles County Museum of Natural History). Their kindness and patience were important to my early career. But, as will be revealed in this book, seed-shrimps led me into a very unexpected and different subject – classical optics.

Michael Land (Sussex University), Sir Eric Denton (Marine Biological Association of the UK, Plymouth) and Peter Herring (Southampton Oceanography Centre) in England had produced some inspiring work on optics and colour in animals. It was great to join in their subject, and I thank them for all the help they gave me, and for tolerating my strange enquiries. After training in the subject of animal structural colours I was ready to bother the optical physicists, particularly Ross McPhedran and David McKenzie (following a significant introduction by their colleague, Maryanne Large) at Sydney University (although many others gave considerable time to my cause). Thanks to these physicists I quickly became familiar with an otherwise unfamiliar subject from its beginnings. And I have found the application of optics to nature quite fascinating.

Looking forwards, sideways, or who knows which direction, I caught a glimpse of the Cambrian. I was steered around the subject of Cambrian biology by numerous palaeontologists. In particular I am grateful to Greg Edgecombe (Australian Museum), Simon Conway Morris (Cambridge University) and the late Stephen Jay Gould (Harvard University) for thought-provoking discussions and comments on my work, and Des Collins (Royal Ontario Museum, Toronto,

Canada) for the trip of a lifetime to the famed 'Burgess Shale' quarry in the Canadian Rockies.

Many of the above people supported my move to Oxford University, and I thank Marian Dawkins and Paul Harvey for making that possible. And then there is the small matter of funding, without which my research would never have begun. This commenced with research grants from the Australian Museum, Macquarie University and the Smithsonian Institution. Then came more substantial funding (for three-year projects) from the Australian Biological Research Study to examine seed-shrimp diversity, and from the Australian Research Council to investigate structural colours in animals. Today I am fortunate to hold a Royal Society University Research Fellowship, which frees maximum time for research. That has been a huge help, but has been gratefully topped up with grants from the Engineering and Physical Sciences Research Council and the Natural Environment Research Council in the UK. Also I am thankful to Somerville College, Oxford, for making me a Research Fellow as supported by the Ernest Cook Research Fund.

Outside my research career, I have people to thank for their necessary help with this book specifically. Cathy Kennedy, of the Oxford University Press, taught me the trade of writing for an audience beyond that of my academic peers, and must have been horrified by my first attempts – after strict scientific conditioning, the popularisation of science is not easy! Peter Robinson of the Curtis Brown literary agency in London helped to refine my technique. But it was the editors I worked with, particularly Andrew Gordon in the UK (and Amanda Cook in the US), who after struggling through early drafts of half-science-half-popular-science, finally transformed my ideas into something readable. And I thank Jeremy Day of Day & Co., London, and the American scientist Ronald Watts for sparking Chapter 10, which may not have happened without their stimulating discussions and interest in my Cambrian ideas.

Finally I thank my parents, other members of my family and a close friend for their continual encouragement and support of my research career.

1

Evolution's Big Bang

The explosive evolution during the Cambrian . . . one of the most enigmatic episodes in the history of life

DEREK BRIGGS, DOUGLAS ERWIN AND FREDERICK COLLIER (1994)

The 'Cambrian explosion' . . . a pivotal moment in the history of life

STEPHEN JAY GOULD, *Wonderful Life* (1989)

Why was there a radiation in the Cambrian? Our most sincere answer is that we do not know

JAN BERGSTRÖM (1993)

Life as we know it

I have a clear memory of animal diversity classes as an undergraduate. Each week I would open my vintage textbook at a different chapter to find a meaningless black and white line drawing of a representative from a new animal group, blending naturally into its background of page creases, ink blots and previous students' scribbles. All in all, the illustrations were hardly more exciting than the thick, blotted stamps of the antediluvian typewriter. They bore no relation to living creatures, nor could one separate the extinct from the living.

A few years later I lowered my head under water in anticipatory awe of one of the world's natural wonders. All I saw was a dark brown cloud. I had come too close to a cuttlefish for its liking. But as the ink disappeared I adapted to the blaze of colours that strike the eye from every direction. The vast diversity of life forms quickly became apparent in the shallow waters of Australia's Great Barrier Reef. Following my student experiences, I was wholly unprepared for my second introduction to animal diversity. The antlers, domes, fans, brains and pipes of corals were the first to manifest themselves. Polyps, each only a few millimetres across, are the living parts of corals which stretch out their tentacles to feed at night, appearing like small anemones or even upside-down jellyfish. Their hard, supporting limestone structures stretch for over a thousand miles, forming the foundations of this famous reef that is visible even from the moon.

Regardless of their external appearance and lifestyles, corals, anemones and jellyfish actually belong to the same higher classification of animals, known as a phylum (plural phyla) because they share the same internal body plan. That is, the organisation of their internal parts – the nutrient processing factories and oxygen transport systems – is similar. Back in the Great Barrier Reef, the complete spectrum of colours present among the corals was paralleled by an almost complete anthology of animal phyla. So began a journey into the unknown. The coral skeleton of the reef was decked out with gardens of sponges, which matched the corals in their diversity of shapes and colours. The sponges provided shelter within their water-filled passageways for animals belonging to other phyla. These lodgers include the bristle worms – a common group of animals that make up a phylum with earthworms and leeches. Some display shimmering opalescent or iridescent colours, like the bizarre-looking sea mouse, a worm whose appearance is best described as a hedgehog with the iridescence of a compact disc.

Sea gooseberries look like transparent variants of their fruit namesakes, flashing with eight iridescent bands. These alien-like blobs of jelly have an internal body plan like no other group of animals and so belong to a phylum of their own – the comb jellies. Starfish are not only obvious during the day but some glow at night with their bioluminescence, emerging from darkness like an extraterrestrial visitor. Starfish are related to common sea urchins and belong to the same phylum of animals. Giant clams display fluorescent blues, greens and purples.

KINGDOM: Animalia
PHYLUM: Arthropoda
SUBPHYLUM: Crustacea
CLASS: Malacostraca
ORDER: Isopoda
SUBORDER: Oniscoidea
FAMILY: Porcellionidae
GENUS: Porcellio
SPECIES: scaber

Figure 1.1 The division of life into categories of different levels, using the woodlouse *Porcellio scaber* as an example. There are thirty-eight phyla of multicelled animals.

They belong to the mollusc phylum along with another animal rather more infamous for its colour – the blue-ringed octopus. During aggressive spells, the blue rings of this small octopus light up to warn of its deadly venom. The less familiar 'moss animals' live in colonies often with unusual shapes and colours, sometimes appearing like the mosses or lichens found on terrestrial rocks. Worms are ubiquitous but hide a plethora of phyla, such as the 'ribbons', 'peanuts', 'arrows', 'acorns' and flatworms. Ribbon worms, as their name suggests, are ribbon-like in appearance and seem quite placid until they make their presence known with their powerful jaws. Peanut worms are less dangerous and have a swollen rear end. Its similarity to a peanut is questionable, but a brownish colour is the norm. The acorn analogy is even less convincing, although arrow worms are more appropriately named. Similarly, the flatworms are flat, and some of those capable of swimming by undulating their bodies possess colours that can shock.

Although very few insects are found in the sea, the crustacean representatives from the arthropod phylum are often at their most spectacular on the Great Barrier Reef, and include the crabs, lobsters and shrimps. Another phylum that is best known for its terrestrial members is the Chordata. This name may sound familiar because it is the group containing amphibians, reptiles, birds and mammals, including humans. But the fishes of the reef, along with some lesser known animals such as sea squirts and lancelets, also belong to this phylum and were once its only members.

Before leaving the water I found, in precisely the same place, the ink culprit, with about thirty of its comrades. The cuttlefish from the mollusc phylum formed an exact arc around me, tentacles to face, eye to eye. Their brown bodies instantaneously bleached as I moved towards them and they all retreated by precisely the same distance. Then their bodies displayed a wave of colour changes. Brown and white synchronised undulations rapidly flowed along the length of their bodies, then suddenly a 'loud' red cut into the sequence, followed by a calming green as I retreated. Meanwhile, the regions housing their eyes remained silver, like mirrors.

Understanding the variety of life

The cuttlefish eye shows strong similarities to the human eye. This is an example of the evolutionary biologists' red herring – convergence. From similar basic building materials a comparable organ has evolved independently to achieve the same function, in two different phyla. But we have learnt it is the internal organisation of an animal that defines its phylum, not its external appearance. As we saw with the worms, the worm-like shape is shared by a number of phyla, but these are unrelated because their internal constructions are very different. If a worm has a mouth but no anus it belongs to the flatworm phylum. Acorn worms are blessed not only with an anus but also a brain and, of importance, a pharynx (the front end of the gut). We also possess an anus, brain and pharynx, but not the body shape of a worm. Now we can divide the body of any animal into two parts – the innards and the outer layers (the 'skin' and 'shells').

The job of an evolutionary biologist is to make sense of the conflicting diversity of form – there is not always a relationship between internal and external parts. Early in the history of the subject, it became obvious that internal organisations were generally more important to the higher classification of animals than are external shapes. The internal organisation puts general restrictions on how an animal can exchange gases, obtain nutrients and reproduce. So we are more closely related to acorn worms than to flatworms. Also, acorn worms are more closely related to us than to flatworms. The complexity of an individual's development from embryo to adult mirrors the sophistication of internal organisation of the adult. To construct an animal with a complex but specific internal organisation from a collection of just a few cells, a specific method of development is required. As one can envisage, from a few cells more steps are required to form a human baby with all its internal complexity than a simple jellyfish – an infolded ball of three tissue layers. Now we can examine the reason why internal organisations carry so much weight in animal classification. It is worth taking the time to understand this subject since it forms the backbone of evolution.

The internal organisations, methods of development from embryo to adult and external shapes of animals are governed by their genes, the set of instructions carried by the chromosomes within the cells. Copious genes govern internal organisation and development. In contrast, the external shape of an animal is generally under the control of considerably fewer genes. But what governs the genes themselves? First we need to take another look at convergence – similarities in external shapes between animals with different internal organisations.

By external parts of animals I refer to the materials, colours and shapes of the outer layers. These have a closer association with the environment than do internal organisations. The environment includes physical factors, such as temperature and light conditions, and biological factors, such as the animal neighbours. The external parts of an animal, in particular, must be adapted to its specific environment, and they may do so within broad limits set by the internal body plan. If two animals live in the same type of environment, they may share comparable external parts, regardless of their internal organisations. This is possible because the external parts are controlled by a relatively small number of genes, and the chances of those genes mutating to code for the same structures in different species are not remote. If we roll two dice, the chances of both landing on a six are 36 to 1. Even though many more than two genetic mutations will be involved in the evolution of external body parts, single mutations can be retained and accumulated. Consequently if a lamp shell and a razor shell, which belong to different phyla, live on the same type of sand into which they burrow, but also require protection from the same predators, it is not surprising that they share a similar external shape – possibly an optimal design. But their internal organisations remain very different. Internal organisations are under the control of many more genes, which *all* have to mutate at the same time to initiate a new internal body plan. Unlike external architectures, internal body plans cannot be built up gradually because usually they can't function in intermediate stages. This is a monumental difference between the mechanisms that control internal body plans and external parts. A spine on the outside of an animal can begin as a small bump, then pass through intermediate stages from a large bump to a long, pointed spine. Importantly, all intermediate stages can exist in their own right because they provide some advantage for their host. But for a change in body plan that involves the abrupt appearance of blood space, or a sudden flipping upside-down of everything internal, for example, there can be no intermediate stages. Internal body plans cannot be constructed stepwise, and so are less influenced by the environment. Hence convergence of internal body plans does not occur. If we roll a thousand dice, the chances of them all landing on a six are 1,000,000,000,000,000,000 to 1 – extremely improbable.

Charles Darwin and Alfred Russel Wallace were first to realise that evolution, an ever-branching process, is the mechanism responsible for animal diversity. Because modifications in the physical and biological environments are taking place continuously, species must also change continuously to maintain an optimal design (or as near as possible to it). This is adaptation. So a modification in the environment can be thought of as a pressure on the local animals to change. Hence the term 'selection pressure' was introduced.

A minor selection pressure may result in a slight modification in a local animal. An animal walking on the sea floor may develop slightly broader feet to prevent it from sinking if the sand or mud becomes softer. A weighty selection pressure may result in a considerable modification in a local animal. The introduction of a new food source may lead to the evolution of new mouthparts and limbs for movement. A collection of modifications in a population can lead to a new species, all within a single phylum. The fewer the modifications between species, the closer their evolutionary relationship or branching point on the evolutionary tree. Here I have been talking about external characters only. Animal phyla today have unique internal organisations, and a mixture of unique and shared (convergent) external characters. But did their internal organisations evolve *in tandem* with their characteristic shapes? And *when* did these both evolve? These questions lead us to the major evolutionary problem that this book will attempt to solve. They will be asked again a little later in this chapter when, after an exploration of the history of life on Earth, they will be easier to digest.

The Cambrian explosion in brief

Thirty-eight animal phyla have evolved on Earth. So only thirty-eight monumental genetic events have taken place, resulting in thirty-eight different internal organisations. Members of these phyla possess a variety of appearances – or external forms – as we have explored on the Great Barrier Reef. Think of the protective spines, swimming paddles, burrowing shapes, grasping arms, eyes and colours. We have also seen that sometimes the same forms can occur in members from different phyla (convergence), but in general each phylum contains a characteristic variety of external forms.

The first fossils from the time 543 to 490 million years ago were found in the Cambrian Hills in Wales. Hence this period became known as the 'Cambrian' (as named by the great Cambridge geologist Adam Sedgwick). It follows that the time span prior to 543 million years ago is called the Precambrian (the Precambrian can be further divided). What if I stated that, based on external characters, 544 million years ago there were perhaps three phyla? Most people would picture a scenario where the number of phyla simply increased gradually from three to thirty-eight over the past 544 million years. Along this trail of thought, 320 million years ago there might have been some twenty distinguishable phyla. Such a steady progression involves a type of process known as 'micro-evolution'. Darwin and Wallace thought along these lines.

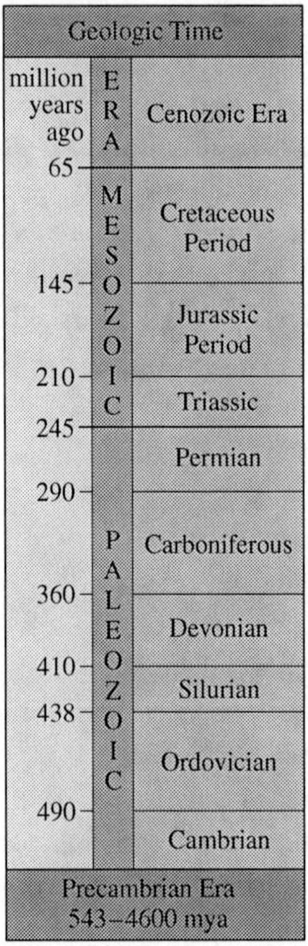

Geologic Time		
million years ago	ERA	Cenozoic Era
65	MESOZOIC	Cretaceous Period
145		Jurassic Period
210		Triassic
245	PALEOZOIC	Permian
290		Carboniferous
360		Devonian
410		Silurian
438		Ordovician
490		Cambrian
Precambrian Era 543–4600 mya		

Figure 1.2 The geological timescale and epochs.

Revolutions in evolutionary theory have occurred since Darwin's time. Now we know that the history of life on Earth has been dominated by long periods of gradual evolution – 'micro-evolution' – or even a complete standstill. But these periods ended abruptly as they were replaced by 'macro-evolution' – short but prolific bursts in evolutionary activity, hence a so-called 'punctuated equilibrium' model for evolutionary history. Darwin and others of his time cannot be blamed for overlooking macro-evolution because its discovery was a consequence of twentieth-century fossil finds and the development of modern biochemical techniques, encompassing genetics and the biology of development from embryos to adults. Events that cause macro-evolution include a faster development from embryo to adult form, the development of sexuality in juvenile forms, and the turning on or off of *major* genes.

With all this in mind, I should like to change the facts and state that 544 million years ago there were indeed three animal phyla with their variety of external forms, but at 538 million years ago there were thirty-eight, the same number that exists today.* In this case the vast diversity of body architectures observed on the Great Barrier Reef would all have appeared during a five-million-year interval (some researchers say fifteen), beginning 543 million years ago. In fact such an interpretation is closer to the truth, and this particular five-million-year interval hosts the subject matter of this book – the 'Cambrian explosion'. The Cambrian explosion is the evolutionary episode in which all animal phyla attained complex external forms. In other words, it is the event during which animal phyla changed from all looking the same to looking different. Now that I have introduced the Cambrian explosion, can I end the first chapter of this book here? Unfortunately not. Such a simple description of the spectacular transition in evolution from Precambrian to Cambrian times does not provide a fair description of how today's diversity of life came into being. We cannot consider only the external appearances of animals but need also to think of their internal body plans. To understand what the Cambrian explosion *really* is, this is essential. Previous explanations of the Cambrian explosion have been greatly simplified by the definition 'the sudden evolution of all animal phyla'. This flippant approach to the most dramatic event in the history of life is misleading in the extreme, and has led to a number of false explanations for the cause of the event. The crux of the problem here is that internal body plans and external parts have been treated collectively, and their evolution is thought to have occurred simultaneously. This is not true. The Cambrian explosion is all about external body parts only. But we have learnt of the great significance of internal body organisations to animal diversity and should study this

*With the exception of one or two extinctions.

subject further if only to provide the outside pieces of the jigsaw puzzle to be solved in this book – *what caused the Cambrian explosion?* The story of internal body plan history takes us deep into the Precambrian.

Up till now we have been measuring time in units of millions if not billions of years. Such quantities are hard for us to make sense of. We think of ancient history as perhaps a couple of thousand years ago. Ten thousand years would be extremely difficult to conceptualise, a hundred thousand, let alone a million, inconceivable. So hundreds of millions of years of evolution are way beyond the realms of the most vivid human imagination. If it is of any help, I began to conceptualise one million years after seeing the immense valleys in Hawaii that have been formed by one million years of running water. These perfectly triangular valleys that terminate at the coast are over 100 metres deep. But a million years ago they did not exist, and the north-west coast of Hawaii had a continuous cliff face with a flat top. As volcanoes formed inland, so did streams or small rivers terminating at the coast. The action of this running water gradually wore a groove into the surface of the ground. And over a million years, water can form a groove 100 metres deep – this is worth thinking about. During such a time period, and without taking space into account, even outcomes with almost negligible odds can emerge. But only when the process in question can change gradually, where each step or small change is saved, and the process can then proceed from a new starting point with this change firmly in place. This line of thought will be continued in this chapter with Sir Andrew Huxley's criticism of a 'jumbo jet in a junkyard'.

'The History of Life' from the very beginning

A book on the complete 'History of Life' on Earth would comprise tens of chapters, where the real subject of *this* book, the Cambrian explosion, would occupy chapter nine in the thesis, as I will go on to explain. A summary of the chapters following chapter nine, such as those where dinosaurs first appeared, and then disappeared, would offer no help in comprehending the Cambrian explosion. But one would be somewhat lost in a book that began with chapter nine, even though it may be the most alluring, without a summary of the previous chapters. Also, I have mentioned that we would look back in the Precambrian at the story of internal body plans. So before returning to the biggest macro-evolutionary event of all, beginning 543 million years ago, I will attempt to paint a word picture of the world as it was then and before that time.

The fine details of life's earliest history, or first chapters, are more open to debate than those of the last 550 million years. This is partly because the fossils that we have from the more ancient times are either microscopic or preserved in poor detail. But it is also because the further back in time we go, the more different the environment was from what it is today and hence the greater the inaccuracy of any extrapolations we may make. Since chapters one to three in 'The History of Life' deal with the longest periods of time, they will be considered in the most detail.

The Earth formed some 4,600 million years ago, and it is generally accepted that life came into existence around 3,900 million years ago, following a flurry of meteorite bombardment. But during the first 3,000 million years of life's history, or chapters one to three, the Earth was populated only by bacteria, algae and single-celled animals. The history of life is written into the Earth's rocks as fossils or preserved in primitive environments. To investigate the first chapters of Earth's story, or the first stages in evolution, we must visit hot, volcanic pools or go deep into the ocean.

Chapters 1 to 3 in 'The History of Life' – the first cells

Thousands of metres below the ocean surface today, black smoke pours into the water from the submarine ridge known as the Axial Seamount, 300 miles west of the coast of Oregon. As dramatic flashes of colour and air flare from the primeval cauldrons or chimneys known as hydrothermal vents, or 'black smokers', one can really begin to form images of a very primitive Earth. There is justification in this imagery because black smokers would have emerged with the appearance of the first seas. They mark the separation of boundaries between Earth's massive plates, on which we live, that float on the planet's surface. Up through the gaps created, hot magma oozes out of the Earth's crust to form new sea floor. The unstable concoction of chemicals ejected from the first black smokers reacted with seawater and provided conditions that could have given rise to the inorganic construction of amino acids and other prebiotic organic molecules that are the building blocks for life. Such chemical reactions compare with those found in primitive living creatures today. As one can imagine, chemicals leaving a black smoker are hot. But very primitive bacteria can tolerate temperatures of up to 110° Celcius in today's black smokers, so heat was never a problem for early life. In fact the living representatives of *all* of life's most primitive species require very hot temperatures to sustain their chemical workings. It is also interesting that black smokers are probably the only places on Earth where the energy of life is not drawn from the sun by means of photosynthesis in an oxygenated atmosphere. The small iron sulphide globules found in the chimneys of the black smokers quite possibly provided the reducing environment necessary to sustain the first life forms. All things considered, black smokers are good candidates for the cradle of life on Earth, and belong in chapter one of 'The History of Life'.

The Murchison meteorite that hit Australia in 1969 contains around seventy-four amino acids, and at least eight of these are of the type that makes up proteins. Could life on Earth have an extraterrestrial origin? Current evidence suggests not. Space is full of organic molecules (those that make up living organisms, including amino acids). But the concentrations they form on impact with Earth, such as in interplanetary dust within the ocean, are much too low to induce life. Hoyle and Wickramasinghe, advocates of the outer space origin, once calculated that the chances of life starting on Earth independently, within a watery soup of amino acids, were roughly the same as having a junkyard spontaneously forming a jumbo jet. Andrew Huxley was among those who set the record straight by explaining a jumbo oversight in these equations. Hoyle and Wickramasinghe had calculated the probability that the right amino acids would come together by chance, in the right order, to form an active protein molecule. But they admit to leaving out two absolutely enormous factors – time and space. The calculation is reasonable enough, but the answer it gives is the probability of this protein molecule originating spontaneously at one particular moment in time and at one particular point in the Earth's oceans. Huxley pointed out that this is of no interest to anyone; what we are concerned with is the chance of a primitive living system being formed at *any* moment within a period of hundreds of millions of years, and at *any* point within the enormous volume of the oceans. In fact the idea of omitting these enormous factors would have been unbelievable if they had not actually done it! Hoyle and Wickramasinghe also assumed that two thousand active protein molecules have to be formed simultaneously by chance to make a primitive living system. But once the first protein molecule had assembled, a self-replicating system would have come into play, which would then develop by natural selection. As Stanley Miller, who first attempted to simulate the origin of life in the laboratory, famously stated, 'The origin of life is the origin of evolution.'

As expected for such a fundamental question, there are many explanations for the first stage of life on Earth, although most researchers now agree that a hot region was involved. Today, American scientists are investigating the problem at the undersea hot springs in the Pacific Ocean. The ocean is not a theoretical necessity. Suitable heat exists in the ground waters deep in the Artesian Basin in Western Australia. And the 'hot' water that exits under ground in the volcanic regions of the USA accommodates a spectacular possibility for the second stage in evolution – chapter two in 'The History of Life'.

Just a couple of thousand metres beneath the surface at Hawaii's Volcano National Park and parts of Wyoming's Yellowstone National Park, there is molten rock that heats the rocks on the surface. Ground water consequently boils in some regions and flows up channels through rocks until it either bursts from the surface as geysers and vapour, or collects to form steaming pools. On its journey to the surface, the water collects minerals from the circumventing rocks. Together with those gained from the molten rock, the minerals become concentrated in the steaming pools, or are deposited as surface waters evaporate. A range of colours can be seen in and around the surface waters, and these belong to colonies of bacteria, which flourish on the minerals. These bacteria represent the second stage in life's history, for they do not have to rely on the finite organic compounds that would have originally accumulated in the Earth's bodies of water. Instead they make their own organic compounds within their cell walls, drawing energy from sunlight. This process is photosynthesis, and requires hydrogen. The bacteria obtain their hydrogen from hydrogen sulphide ('rotten eggs' gas), originating from an underground reaction between the molten rock and ground water. But such a delicately balanced diet means that these bacteria are restricted to regions of volcanic action. The third stage in evolution had further repercussions: it opened the floodgates for endless possibilities of life forms.

Chapter three in 'The History of Life' sees the appearance of cyanobacteria (traditionally and erroneously called 'blue-green algae'), organisms that obtain their hydrogen from water. This was achieved by the evolution of a substance of great consequence – chlorophyll, the lifeblood of true algae and the higher plants. Unlike hydrogen sulphide, there is an extensive supply of water on the planet, and this equates with a profuse occupation of Earth by life. As the cyanobacteria removed hydrogen from the Earth's water, oxygen remained and entered the atmosphere. Cyanobacteria include the simplest forms of life existing today, and the timing of the first cyanobacteria is known from fossil evidence.

At Pilbara near Marble Bar in Western Australia, a fine-grained mineral called chirt can be found in rocks 3,500 million years old. Slices of this chirt are cut so thin that they are translucent and can be examined with an ordinary microscope, revealing the shapes of cyanobacteria. But how do we know the fossils really are cyanobacteria like those of today? After all, the organisms are little more than minute squiggles. The answer lies in the large, unique structures that are formed by the micro-organisms, structures that are still formed today.

In the same Australian state as Pilbara, Hamelin Pool can be found within Shark Bay. Here coral reefs are replaced by stromatolites (from the Greek, meaning 'stony carpet'), appearing like large button mushrooms carved from rocks that rise above the shallow sea. The entrance to Hamelin Pool is blocked by a sand bar and eel grass. This barrier separates the water in the pool from the ocean, and the evaporation of water increases the salt concentration of the pool. Animals that usually feed on the cyanobacteria in the pool cannot survive under such salty conditions, and so the cyanobacteria thrive. The cyanobacteria exude lime, which hardens to form the stromatolites. We know that the Pilbara chirt is actually made up of ancient stromatolites because it shares the same unique structure as the stromatolites of Hamelin Pool. Hence the Pilbara chirt comprises the first known tombstones which record the beginnings of life (though there is chemical evidence from Greenland that suggests life was present on Earth 350 million years earlier, but this has yet to be widely accepted). So Hamelin Pool may represent a scene that could have been seen on Earth some 2,000 million years ago. And importantly, thanks to the cyanobacteria, the Earth gained an oxygenated atmosphere around this time. Atmospheric oxygen not only permits breathing in higher animals but also provides a protective barrier – the ozone layer – from the sun's ultraviolet rays, which can be harmful to animal tissue.

A long period in the history of the Earth followed where, as far as we know, nothing of any great significance happened. But, and just as mysterious, came another huge step, or chapter four in 'The History of Life' – the appearance of cells with a nucleus.

Chapters 4 and 5 in 'The History of Life' – the nucleus and the grouping of cells

The organisms found in the first three chapters of life's history book are single-celled and have their DNA distributed irregularly throughout their cells. The new organisms to appear are also single-celled but have a distinct nucleus packed with DNA and separated from the watery fluid of the cell by a membrane. Outside the nucleus there are other units such as mitochondria, that produce energy for the cell by using oxygen in a similar manner to bacteria. The nucleus is the main organising force of the cell. The first cells with a nucleus appeared around 1,200 million years ago and belonged to a group of single-celled organisms called protists. There are around 10,000 species of protists today, including the familiar amoeba. Protists can be seen readily when a drop of pond water is viewed under a microscope. Some possess a thrashing tail or fine rhythmically beating hairs, while others contain packets of chlorophyll that, like cyanobacteria, use the energy of sunlight to produce

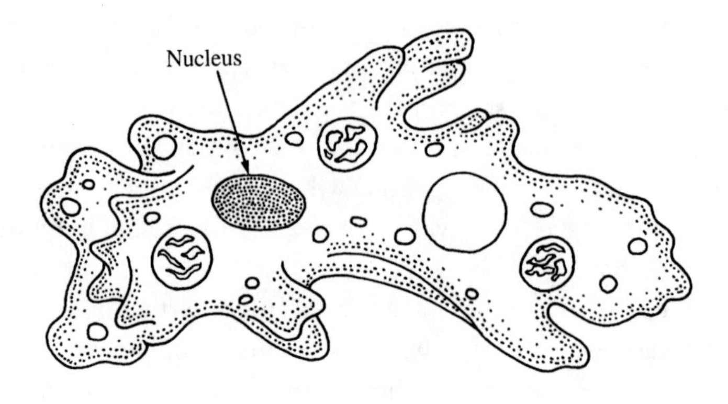

Figure 1.3 An amoeba – a cell with a nucleus and organelles.

food for the cell. These packets of chlorophyll and the mitochondria have their own DNA. Some researchers believe that the cells with a nucleus are the combination of a number of cells without a nucleus, each performing a specific function to maintain a life system.

Protists reproduce by splitting into two, an action known as binary fission. But there is much more to the binary fission of protists than there is to the binary fission of bacteria because, unlike in bacteria, in protists most of the separate internal structures have to split. The DNA of the nucleus divides itself in a particularly intricate manner so that its genes are copied and one complete set is passed to each daughter cell. Although methods vary within the group, the key feature of nucleated cell reproduction is that genes are shuffled around. One of the mechanisms employed involves two cell types: an egg and a sperm. This is the origin of sexuality. Here, genes are distributed to daughter cells from two parents rather than one. The daughter gene sets reflect the new combinations of parent genes, and occasionally these new combinations are so divergent that they produce a slightly different organism with new characteristics. This is another form of evolution. And with the establishment of sexuality, the possibilities for genetic variation increased and evolution accelerated.

There is a limit to the size of a single-celled organism. As the cell becomes larger, the internal chemical processes become less efficient, and eventually reach a point where the organism is no longer viable. The next step in evolution, occurring in chapter five in 'The History of Life', was to bypass this limitation by grouping cells together in an organised colony. *Volvox* is a species that has done exactly this. *Volvox* is a hollow sphere, about a millimetre in diameter, where the wall is made up of cells, each with a rhythmically beating hair appearing like a tail. The movement of the hairs is coordinated to move the entire sphere in one direction. The next group of cells to evolve had an additional character – a cuticular stalk that is branched to unite small colonies of cells. But the following, very important, step was the division of labour between the component cells of a colony, around 1,000 million years ago. This step was manifest in the beginning of chapter six in 'The History of Life' – the appearance of the sponges, the first true multicelled animal phylum.

Chapters 6 to 8 in 'The History of Life' – appearance of the true multicelled animals

Sponges have only a few cell types modified to perform specialised functions, and the sort of cell-to-cell junctions that form sheets of tissues in higher forms are absent. In general, sponges have open-topped, sack-like bodies which are fixed to the sea floor. Water is pulled through the body and food is filtered out. They are the only multicelled animals with cells capable of independent survival. If a sponge is passed through a sieve the individual cells separate but continue to survive and even reproduce. Sponges also lack a nervous system and muscle fibre, characters possessed by the next two most derived phyla, Cnidaria (with a silent 'c' – this phylum includes jellyfish, corals and sea anemones) and comb jellies. Cnidarians and comb jellies have two thin but clearly modified tissue layers separated by a gelatinous material. One layer is protective and surrounds the body; the other has a digestive function and forms the lining of a gut. Cnidarians and comb jellies have a basic body plan that is also a sack-like form, but at one end there is a mouth which can be opened and closed and tentacles which direct food to the mouth.

Chapter seven in 'The History of Life' opens with the evolution of a body plan where three primary tissue layers exist but a blood space between tissue layers is absent. The animals with such an internal organisation are the flatworms. Flatworms have an inner tissue layer that produces muscles and some other organs – obviously a layer with a future – but they are without a blood circulatory system. This means that oxygen has to be transported to the inner tissue layer by diffusion, which works very slowly and its efficiency decreases as one thickness of tissue increases. This means that the animals must be flat, which indeed they are. Like jellyfish, flatworms have guts with only one opening, which is a port for both incomings (food) and outgoings (waste). But the evolutionary position of flatworms is uncertain, and so too is the relationship between chapters seven and eight in 'The History of Life'. Controversy aside, in chapter eight of 'The History of Life' the next evolutionary innovation takes place – a body with again three modified layers of tissue but also an open blood space. This is followed, in the same chapter, by the appearance of a body plan with three modified layers of tissue, a blood space in the form of blood vessels *and* an internal body space, in which the gut is suspended. But the appearances of a blood space and a body space were no run-of-the-mill evolutionary innovations. They paved the way for the evolution of the further internal variations that discriminate the remaining thirty-four animal phyla, including arthropods (crabs, insects and spiders), molluscs (snails and squid), echinoderms (starfish and sea urchins), chordates (fish and mammals) and many other weird and wonderful phyla that have not made household names. The obvious question to be posed is: 'When exactly did the step take place from about three to thirty-eight phyla, still within chapter eight in "The History of Life"?' The similar question posed earlier in this chapter has now become refined and more understandable. But before attempting to answer this new enquiry, we should take a moment to pause and reconsider what life's history book has taught us so far. It is important to remember that this question does not refer to the Cambrian explosion, but rather to prior events.

We know that phyla are defined by internal body plans, and we have now reached a stage in life's story where all thirty-eight body plans of multicelled animals are in place on Earth. But we have not yet

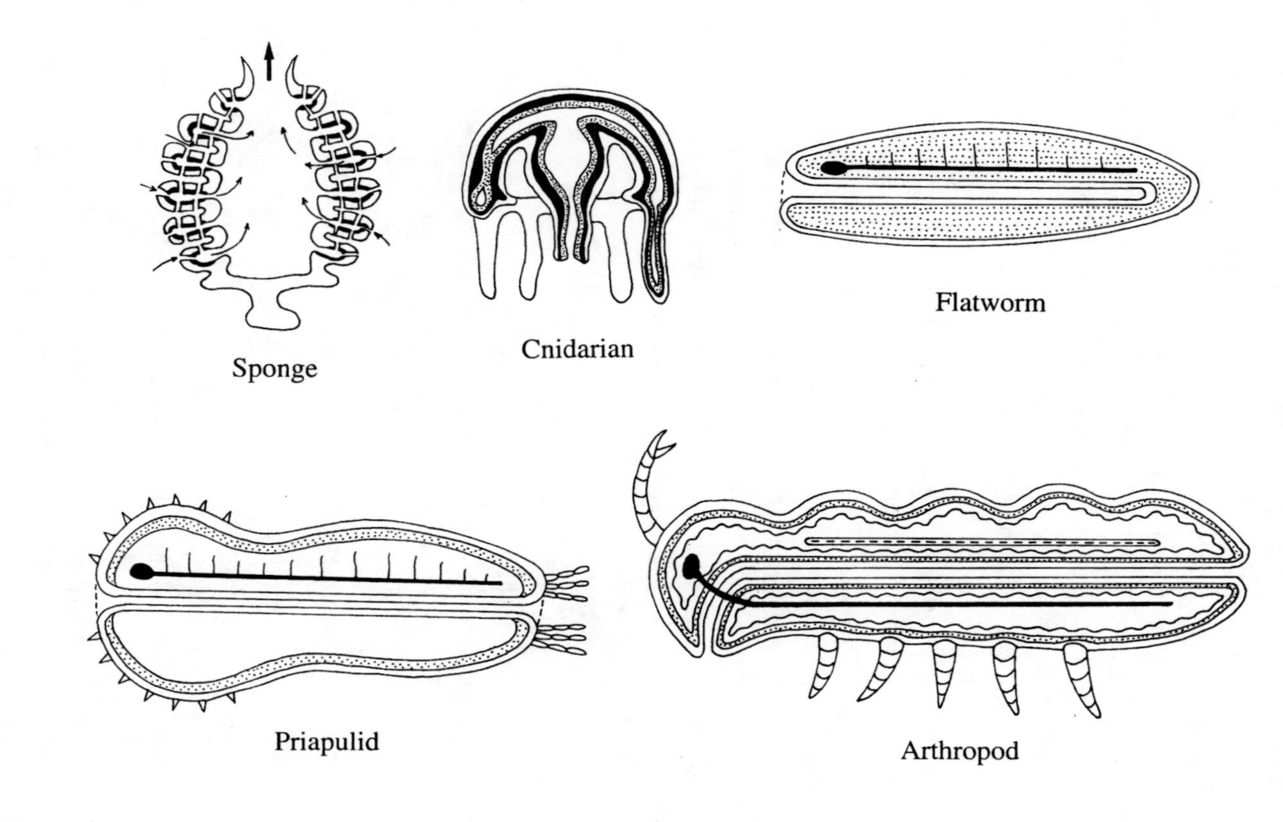

Figure 1.4 Sections through representative bodies of different phyla showing simplified examples of internal body plans.

considered the external appearances of these animals. The most advanced animal whose shape has been considered up till now is the comb jelly, or possibly the flatworm. So what we have in chapter eight are three primitive phyla – sponges, jellyfish and comb jellies – with their own distinctive body plans *and* body shapes, and a bunch of worm-shaped, or soft-bodied forms, each with one of thirty-five different internal body plans, including that of the flatworms. Is this picture accurate? Were the internal body plans of crabs and starfish really once hidden within the soft body of a worm? This 'all-worm' scenario does not seem so far-fetched when we consider that many different phyla still possess a worm-like body today. Remember the ribbon, peanut, arrow and acorn worm phyla? Also we know that the most primitive forms of some phyla, including the chordates to which we belong, had the shape of a worm. But this is far from conclusive evidence. If the 'all-worm' scenario is correct, we are faced with a chapter nine in 'The History of Life' that deals with the evolution of external body forms, leading on from a chapter eight where only the internal body plans of phyla are in place. What *does* chapter nine have to say? At what points in geological time does it begin and end? Using their genetic dating techniques, or molecular clocks, the biologists tell us that the internal body plans of all phyla evolved between 1,000 and 660 million years ago, in chapter eight of life's history book. To learn about external body history and the Cambrian explosion making the intrepid leap to chapter nine, we must turn to the fossil record.

Chapter 8 in 'The History of Life', continued – the Ediacaran enigma

The Flinders and Mount Lofty Ranges are dominant features of the state of South Australia. They extend like a backbone from the coast, near Adelaide, to distant inland regions. These ranges became the subject of some renowned geological study, which resulted in a thorough explanation of the eventful geological history of the area.

Sediments were deposited into an elongated trough in the ranges, and a sequence of rocks 24 kilometres thick gradually accumulated. When the stresses in the Earth's crust subsequently changed, the entire mass of sediment was folded and pushed up to form a predecessor of the present ranges. Some of the oldest known cells with a nucleus, in addition to stromatolites up to 1,600 million years old, have been found fossilised in this sediment. But more importantly, sometime between 1,400 and 900 million years ago a sandy beach developed on the pre-existing crystalline rocks that had formed a coastal trough. Fortunately for palaeontologists, the sandy beach environment continued into the Cambrian period, up until 540 million years ago.

In 1947 Australian geologist Reginald Spriggs collected fossils of multicelled animals from the Ediacaran Hills in the Flinders Ranges. These fossils were from the Late Precambrian epoch, about 570 million years ago. But because everyone 'knew' there could be no fossils from this geological period, Spriggs' professor duly placed the rocks next to the dustbin. Spriggs' enthusiasm got the better of him and he rescued his fossils to give them closer inspection and reprieve from an undignified end. Such an end would have been inappropriate for the spoils of Spriggs' labour, which has resulted in the universal term 'Ediacaran fauna'. This is the name given to collections of the earliest known multicelled animals, the first of which was found in Spriggs' enigmatic rocks. Although the Australian site has yielded the greatest variety of Ediacaran organisms, they have since been discovered in Africa, Russia, England, Sweden and the USA. The oldest Ediacaran fossils derive from the remote Mackenzie Mountains in Canada's Northwest Territories. These impressions are interpreted as soft, cup-shaped animals that lived on a muddy sea floor around 600 million years ago. They have become accepted as the oldest known multicelled animal fossils in the world.

The first Ediacaran fossils discovered look like flower impressions. These may have been blobs of living matter that were washed up on to a beach, baked in the sun and then covered by a wash of fine sand by the next tide. Walking along the beach on Heron Island, in the Great Barrier Reef, one can find the divided, circular shapes of jellyfish that have become beached and are about to go through a similar process of eternal preservation. The chances are that these too will soon become similar flower-like impressions. Were the Ediacaran organisms jellyfish? Other Ediacaran fossils appear frond- or feather-shaped. Once underwater off Heron Island, similar shapes can be seen waving from the sandy bottom in the form of sea pens that share a phylum with jellyfish. At least sixteen different species of Ediacaran organisms have been identified, but what kind of animals were these ancient creatures? Were there really sea pens and jellyfish among them?

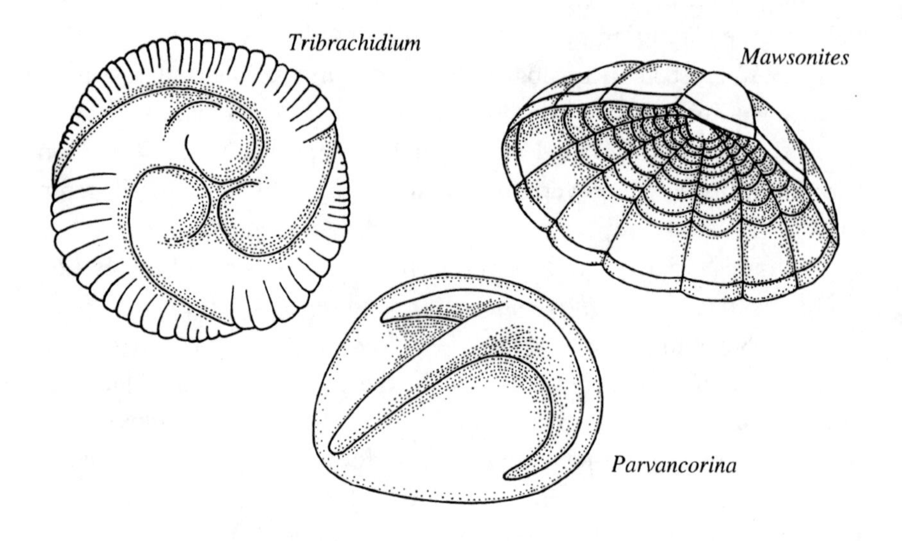

Figure 1.5 The Ediacaran animals *Tribrachidium, Mawsonites* and *Parvancorina.*

At first sight, many of the Ediacaran organisms look like species living today. But some of these similarities may be explained by our old stumbling block – convergence. Take *Dickinsonia*, for instance. The Precambrian species *Dickinsonia* appeared elliptical in its shape from above. It grew to about a metre in length yet was less than 3 millimetres thick. Several hundred specimens of *Dickinsonia* have been collected from the Ediacaran Hills – it must have been quite common. It is also known from northern Russia, and presumably inhabited a large proportion of the Late Precambrian globe. *Dickinsonia* shows dividing lines radiating from a central region to the edges of the animal. If these lines are interpreted as segment-dividers, a body made up of separate, connected segments is conceived. Then *Dickinsonia* could be assigned to the phylum of segmented worms, to which living bristle worms, earthworms and leeches belong. But there is another possibility even if we continue down the segmentation path. If the 'segmentation' of *Dickinsonia* evolved as a means of increasing its body size it could belong to another phylum, because the equivalent character in segmented worms evolved in response to soft layers of sand, to facilitate burrowing. So for what purpose did the 'segments' of *Dickinsonia* evolve? There is more evidence that the fossil record can provide towards this question, but it doesn't involve fossils of the animals themselves.

Despite the number of fossilised specimens known of *Dickinsonia*, no fossilised burrows have been found that could have accommodated this species. Yet burrows are known to preserve as fossils – such marks left by ancient animals are known as trace fossils. But surface trails

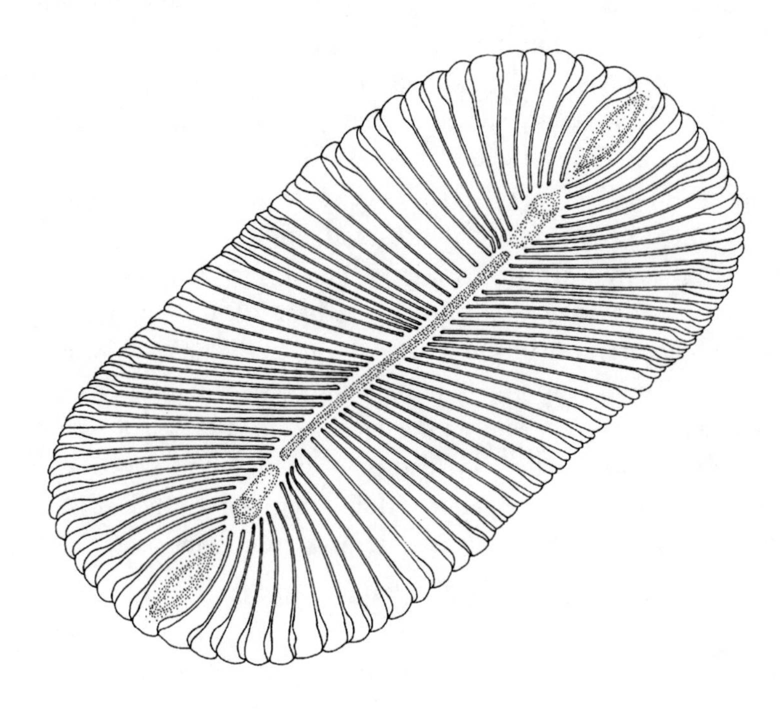

Figure 1.6 The Ediacaran animal *Dickinsonia costata.*

rather than burrows are the characteristic trace fossils of the Ediacaran epoch. This suggests that some Ediacaran animals crawled around on the sea floor and none burrowed into it. Now we can cross segmented worms off the Precambrian list, at least in their segmented forms, with their burrowing lifestyles. So *Dickinsonia* was not a segmented worm.

Recently, researchers have been able to unravel the ancient ancestors of Precambrian animals by comparing the squiggles and swirls that mark their trails with the wriggling trails left by worms and other soft-bodied animals today. Thus, problems that had been unsolved following years of anatomical study are now being resolved.

Signs of internal features of the Ediacaran animals are not evident, but theories relating them to jellyfish and sea pens (cnidarians) have become accepted as serious propositions. Indeed, we have found probable embryos of jellyfish (and sponges) in Chinese rocks 570 million years old. This would place the phylum Cnidaria in the Precambrian with a diversity of external forms or body shapes. Also, it is now believed that many of the Ediacaran fossils represent animals from more derived phyla, albeit before they possessed their characteristic shapes of today. This will be addressed later.

The gap in geological time that once separated the Ediacaran fossils from the next suite of fossils to be found, and once provided evidence that the Ediacaran organisms were a 'failed' first attempt at evolving animals and consequently died out, has been filled. Now it is known that the Ediacaran organisms lived right up to the next major event in animal evolution. Not only that, but the last six million years of their existence appears to have been the period of greatest Ediacaran diversity. No one knows why they died out, although this probably has a good deal to do with the sudden, 'blitzkrieg' appearance of the next dominant forces on Earth. It is time now to consider the Cambrian explosion.

Chapter 9 in 'The History of Life' – the Cambrian explosion

The Cambrian explosion, which post-dated the Ediacaran fauna, is a milestone in evolution that can be matched in significance only by the beginning of life itself. It paved the way for the emergence of the vast diversity of life found today, whether in Australia's Great Barrier Reef or Brazil's tropical rainforests. It involved a burst of creativity, like nothing before or since, in which the blueprints for the external parts of today's animals were mapped out. Animals with teeth and tentacles and claws and jaws suddenly appeared. Explanations for this grand event do not have such a deep history as theories on the origin of life, only because the Cambrian explosion is a recent realisation. Darwin, among others, became puzzled by the sudden appearance of hard-shelled fossils at the beginning of the Cambrian period, about 543 million years ago, and by their apparent lack of evolutionary ancestors. Darwin and his contemporaries hypothesised that early forms of each animal phylum did not fossilise or became entombed in old rocks that were unsuitable to provide preservation as fossils. But as we have seen, Darwin had only micro-evolution to work with. Now we have so many examples of well-preserved sedimentary rocks (suitable for fossil preservation) from before the Cambrian that it is no longer reasonable to claim that conditions only became suitable for preservation during the Cambrian. Today's view of the fossil record invokes a Cambrian 'Big Bang' in the evolution of external body parts from soft, worm-like forms.

The Cambrian is a relatively brief period in the history of the Earth yet outstanding in the history of life. Spanning just forty-three million years it was a period of monumental change. The earliest hard-shelled fossils that Darwin pondered over were later revealed to have appeared even more suddenly. They were narrowed down to the Cambrian period on the discovery of the fossils of the Burgess Shale. These abundant fossils, of the fauna and flora communities that existed 515 million years ago, have been the subject of many spirited scientific discussions and are deserving of their place in the history of science. They have been regarded either as the predecessors of today's fauna and flora or as enigmatic species that belong to phyla that did not survive the Cambrian. We now prefer the interpretation that the Burgess Shale organisms can be accommodated within today's thirty-eight phyla (actually a few of these are extinct today).

The Burgess quarries today

Although barely cutting through the distant haze, transcendental mountains are prominent over the otherwise featureless landscape viewed from Calgary in Alberta, south-western Canada. One is instantly drawn to those geological wonders, the Rocky Mountains, and the attraction becomes stronger as they are approached via the Trans-Canadian Highway. Banff National Park is the first port of call on entry to the Rocky Mountains. Everywhere there are mountains that could be termed spectacular even when compared to any other mountain range. It is the steepness of the mountainsides, their jagged peaks, endless variety of shapes and their 'contour lines' running in conflicting directions that makes this place so unique. The 'contour lines' swirl around the landscape, continuously pulling the eye in different directions and towards different focal points. These lines are actually the boundaries of sediment layers, laid down millions of years ago by sediment in the sea settling out on to the bottom, forming a new sea floor. So although a thousand metres or two above ground today, the rock that constitutes these mountains began its history underwater. As the Earth's plates moved around throughout geological time, and slowly crashed into each other, something had to give. The rocks that now form the Rocky Mountains were one of those things. They were forced up from below the water and into the air, and their chaotic movements produced the uneven patterns of 'contour lines' seen on the mountains today.

Continuing west along the Trans-Canadian Highway, and staying within the Rockies, one enters British Columbia and Yoho National Park. The small mining town of Field lies at the foot of Mount Stephen, famous for its Cambrian trilobite fossils. The rusting iron shacks of Field are gradually being replaced by wooden bungalows and small motels that blend into the coniferous surroundings. This is now a useful base for serious mountain walkers. But there is something unique about Field. Some of the best preserved, complete Burgess Shale fossils are on display at the information centre here. These fossils separate Field from the rest of the Rockies and inspire thoughts that there is something very special about this place. The fossil display is the work of Des Collins, a palaeontologist at the Royal Ontario Museum in Toronto, eastern Canada. Each year Des Collins and his team of helpers and students rent one of the utilitarian wooden bungalows in Field. On my visit to this bungalow in July 2000, I was directed to Des Collins's new and more temporary base camp, at the site of the Burgess Shale quarry.

The Burgess fauna and flora were organisms that lived 515 million years ago in a sunlit marine reef, at a depth of 70 metres or less. More specifically, they inhabited the edge of the reef, at the top of a submarine cliff known as the Cathedral Escarpment. The Cathedral Escarpment probably formed when the edges of the reef became detached and collapsed, sliding several kilometres down the slope. At the base of the sloping Cathedral Escarpment, some 160 metres below the reef, was a basin. One day in the Cambrian, an abrupt inflow of very fine mud swept across the area, burying most of the reef, but not the edge at the top of the Cathedral Escarpment. The Burgess fauna and flora escaped this catastrophe, which saw an end to carbonate deposition on the reef – the carbonates ending up in the basin. But further inflows of fine mud were to follow, and eventually the Burgess fauna and flora were gathered. The mud flowed over the edge of the reef like ash from a volcanic eruption and carried the Burgess organisms down the face of the Cathedral Escarpment, dumping them into the basin. Here they were preserved in all sorts of positions, akin to the bodies entombed at Pompeii. Today the Burgess organisms are found fossilised, albeit flattened, in a block of rock formed from compression of the mud in the basin, above the layer of carbonate. They serve as a snapshot of a community of life that existed in the Cambrian, 515 million years ago.

What have become known as the Burgess quarries are located 5 kilometres north of Field, on Fossil Ridge. Over a million years ago, the block of rock containing the Burgess fossils was transported 160 kilometres by movement in the Earth's crust. If it had remained in its original position, the heat and pressure of movements in the crust at that particular place might well have destroyed the Burgess fossils.

In 1999 I set out to reach Des Collins's camp and the Burgess Shale quarry on a very grey and wet morning, in the hope that the weather would improve. It did not. But the mist actually created an enigmatic atmosphere, which somehow seemed appropriate. I knew something exceptional lay ahead, but at the same time I did not know what to expect.

The steep climb from the base of Whiskey Jack Falls is rewarded with a view through the pine trees of a small lake with the most emerald green of colours. This lake was created by glacial movement, which stirred up minerals into the body of water left in its trail. Although the rest of the path to the Burgess quarries was less steep, it was still uphill all the way, for about three and a half hours. But it was not the slope that caused the most concern, nor was it the mist. It was the snow. Not the depth of snow covering the path, but the fresh prints of bear paws, claws and all, that it had preserved. After recently bumping into a bull elk, I was quite relieved not to meet the maker of these prints during my climb.

The next lake encountered resembled a setting for one of King Arthur's tales. The mist over the green water also covered most of the surrounding pine trees and all signs of a sky. The air was very still and the silence impressive. A very different terrain and signs of life surrounded the path from here. Beaver-like hairy marmots were playing on and around this 'Burgess Trail' path, which cut across an elongated mountainside that included Fossil Ridge. One of the smaller plants on the edge of the path was particularly interesting because it had leaves that were corrugated or concertinaed to give strength to the thin structure – a flat leaf would have collapsed. I will return to these leaves later in the book.

After crossing a couple of ice bridges I could eventually see the blue tents of the Collins camp in the snow, set against a backdrop of one of the larger lakes in the Rockies (appearing an intense emerald green, of course). The camp was surrounded by temporary electric fences, to keep out bears, and there were red stains in the snow. The red stains had nothing to do with bears, but were the collective red-coloured eyespots of single-celled organisms that can inhabit snow. The 'eye' in eyespot is not really appropriate, for these organs can only sense the direction of sunlight – they cannot produce visual images, or 'see'. The distinction between eye and eyespot will become consequential to this book.

Now all that was left to reach the Burgess quarries was a 200-metre scramble up Fossil Ridge from the Burgess Trail path. Three quarries could be seen, but the original, and most prolific, was the Walcott Quarry. The Walcott Quarry was the site of Des Collins's current excavations, the last in a productive series that began in 1982. This quarry takes the form of a terrace, cut a few metres into the mountainside and several metres wide. At the back of the quarry the various layers of sediment are visible, following the removal of snow, as coloured bands. Each layer was once the sea floor in the Cambrian. The back of the quarry is continually extended into the mountain by hammering iron bars vertically into the rock from above; the bars then act as levers to break the rock away. This rock, or shale, takes the form of thin sheets like slate tiles on a roof. It is painstakingly split into thinner and thinner fractions and observed for fossils, which are more reflective than the bare rock. I examined some of the fossils that had just been exposed to air for the first time in 515 million years. There was a lobster-like animal about the size of a hand, with menacing, grasping limbs and bulging eyes, and some smaller creatures with hard shells, the like of which I had never seen before. Even in the field, with the naked eye, the exquisite detail in which the Burgess fossils have preserved is apparent. It is something exceptional for a biologist to see a Burgess fossil at its original site, and this eclipsed all other amazing experiences undergone in the Rocky Mountains. The Burgess quarries are now protected from unauthorised fossil hunters by national law, which is enforced by Parks Canada wardens. This is appropriate since the fossils of the Burgess Shale are of international importance.

Scrambling back down from the quarry to the Burgess Trail path, piles of shale known as taluses are apparent. Although once discarded from the quarry for being empty, fossils are being found in abundance in the Burgess talus following further splitting. In fact taluses are becoming an increasingly useful source of fossils because the quarry itself is drying up. The talus on Fossil Ridge was left collectively by all generations of Burgess excavators, including the very first – Charles Doolittle Walcott.

A century of research

Charles Doolittle Walcott was the head scientist at the US National Museum of Natural History (Smithsonian Institution, Washington, DC) and the world's authority on the Cambrian. From 1907 to 1924 he undertook expeditions, often with his family, to the Rocky Mountains in Yoho National Park for the purpose of collecting Cambrian and earlier fossils. Cambrian trilobites were known from this region. But during an expedition to Fossil Ridge in 1909, Walcott found *Marrella*, *Waptia*, *Naraoia*, *Vauxia* . . . in fact, amazingly, a suite of soft-bodied fossils which included many, at first sight mysterious animal forms. He made sketches of each species in his field book and attempted to make sense of these forms, which he knew shouldn't really be there. Recognising instantly the importance of his finds, Walcott carefully packed each fossil specimen and shipped them down to his base camp by mule.

It is most unusual to find fossils with details of soft parts preserved; most fossil animals are known only from their hard parts, such as the shells of snails. Needless to say, Walcott and his family planned and carried out numerous expeditions to Fossil Ridge. In his field diary, on the day of his first major Burgess Shale discoveries, Walcott wrote unassumingly, 'We found a remarkable group of Phyllopod Crustaceans.' In 1910 Walcott uncovered the 'Phyllopod Bed', the site known today as the Walcott Quarry. This site yielded a previously unimaginable diversity of Cambrian animal forms. More than 65,000 fossils with both hard *and* soft parts preserved were recovered by the Walcott family and dispatched to Washington by the end of 1911. About 170 species of animals (mainly) and plants have been recognised from these Burgess Shale fossils. Walcott described over a hundred of these himself, although the phyla to which he assigned these species later became the focus of considerable controversy. His first instinct was to place the Burgess Shale fauna into animal groups that still exist today. This instinct was apparent in his original statement, where he forced the species found during his initial discovery into the Crustacea (part of the arthropod phylum), to which today's crabs, shrimps and woodlice (slaters) belong. This was a safe bet – less controversial than constructing new phyla, perhaps, which may have been heavier for

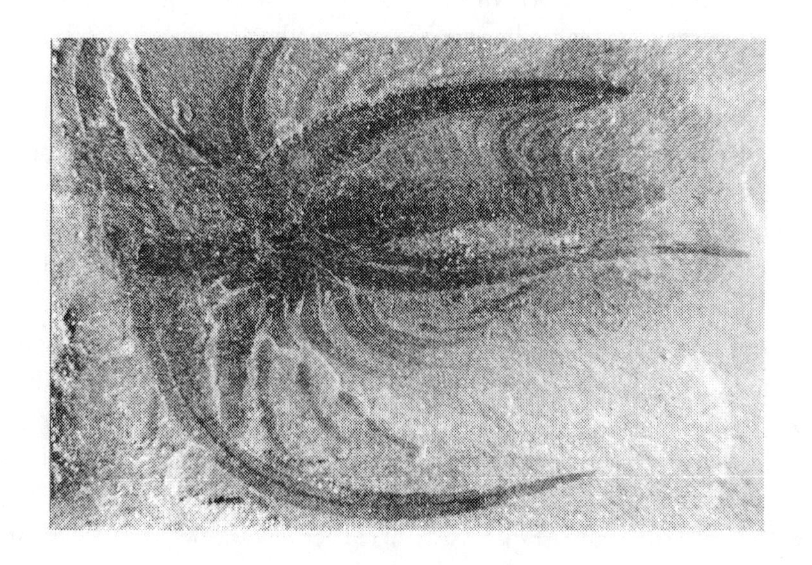

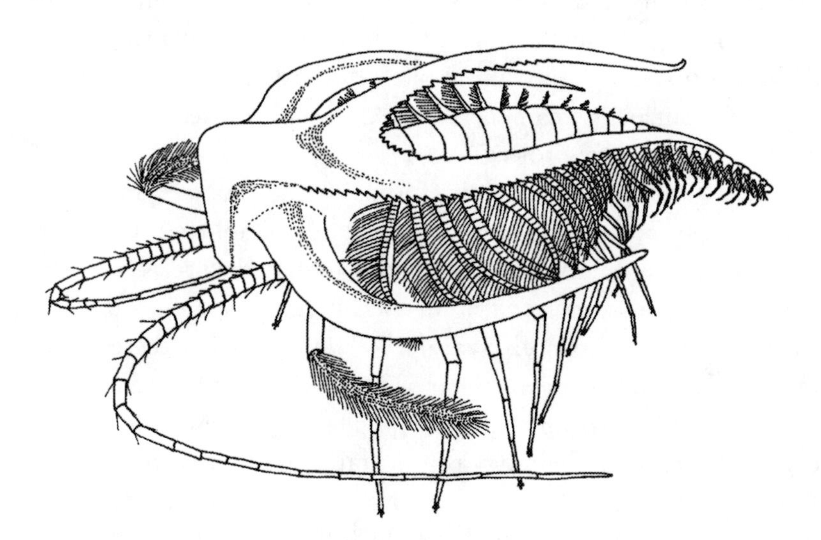

Figure 1.7 *Marrella* from the Burgess Shale – fossil and three-dimensional reconstruction.

Walcott's colleagues to digest. After all, arthropods, with their hard external skeletons, were already known from the Cambrian in the form of trilobites. The use of living phyla continued throughout Walcott's later treatment of his finds. Interestingly, we have gone full circle since Walcott's times. The next wave of scientists felt that many of the Burgess species were deserved of new phyla, but others have since fitted them back comfortably into living phyla, although rather more phyla than those used by Walcott were needed.

From 1924 to 1930 Percy Raymond led Harvard University summer schools to the Canadian Rockies. The Walcott Quarry was visited several times and a second quarry was excavated nearby. From this 'Raymond Quarry' further Cambrian finds were made, though the fossils tended to be less well preserved than those from the Walcott Quarry.

Although Walcott's and Raymond's original accounts received discussion, surprisingly little attention was given to the Burgess Shale fossils until the Italian biologist Alberto Simonetta began redescribing some of the Burgess species, particularly the arthropods, in 1960. Simonetta's work revealed that there was much to gain from a re-examination of the Burgess fossils, and the first significant suggestions that the Burgess animals belong to extinct phyla were made. This introduced controversy to the subject of early multicelled animal evolution, and with controversy came a growth in scientific attention, beginning with the 'Cambridge project'.

Harry Whittington, a world authority on trilobites, initiated the 'Cambridge project' in the 1960s while employed at Harvard University. Whittington initially planned to map the precise levels from where the fossils occurred within the Burgess quarries, a detail neglected by the previous excavators. While exercising these plans he found some new fossils as a bonus. The information Whittington gathered led to an understanding of the original setting of the Burgess Shale organisms and of their environmental and ecological conditions. Workers from the Geological Survey of Canada were chiefly responsible for the environmental findings of the Burgess project, and in 1966 Whittington moved from Harvard to Cambridge University, where much of the major work on the redescription of specimens and ecological aspects of the Burgess ecosystem was carried out. In 1972, Derek Briggs and Simon Conway Morris became involved in the Cambridge project. Originally students of Whittington, Briggs and Conway Morris played major roles in painting a reliable picture of the Burgess ecosystem – the community structure as a whole. This became the earliest ecosystem where the workings are understood in detail. It is one thing to know of an extensive collection of fossils from one particular site, but quite another to understand the ecological workings of the original environment. Because the Burgess environment was, in geological terms, very near to the time of the Cambrian explosion, it had great potential to interest much wider scientific circles. The stage was now set for the next phase of work on the Burgess Shale, which later transpired to be as important as the original scientific investigations.

Work on the Burgess Shale fossils led to the first major understanding of the Cambrian explosion within the community of Cambrian biologists, but for the obscure Burgess animals to attract the attention of a wider audience, and compete in the dinosaur arena, some particularly imaginative and skilful writing was necessary. This first came in the form of Stephen Jay Gould's award-winning book *Wonderful Life*, published in 1989. In his book, Gould succeeded in showing the world that animals once existed on Earth that were far more bizarre than our wildest conceptions of alien life-forms. *Wonderful Life* captured unexpected levels of attention, partly attributed to an ingenious explanation of how we ourselves are involved in the Cambrian explosion. Gould's curtain came down on *Pikaia*, a swimming worm that was the first known member (at that time) of the phylum to which we belong. If *Pikaia* had not survived the Cambrian period, the story goes, then we would not be here today.

Today it is generally believed that ten phyla are represented by the Burgess fauna: sponges, cnidarians (here sea pens and sea anemones), comb jellies, lamp shells, molluscs, hyoliths, priapulid worms, 'bristle worms' (there are also other worms in this phylum), velvet worms, arthropods, echinoderms (here including sea lilies and sea cucumbers) and chordates (to which we belong). Algae and cyanobacteria are also represented in the Burgess Shale biota, along with one or two animals that remain a mystery and have yet to be assigned to a phylum, although this does not necessarily imply that they belong to additional, extinct phyla.

Palaeontological gold

Although the Burgess Shale fauna dominated discussions on Cambrian evolution for many years, other Cambrian assemblages have been more recently discovered. The limestone shale of southern Sweden contains late Cambrian material in stones known as 'Orsten'. This material shows mixed preservation, and includes some complete and exquisitely preserved tiny arthropods such as trilobites and 'seed-shrimps' or their relatives. The Orsten fossils show a type of preservation, called phosphatisation, which is also known from early Cambrian deposits of Comley in Shropshire, England.

The Canadian palaeontologist Nick Butterfield, now at Cambridge University, found Cambrian fossils in borehole samples from Mount Cap, near the Great Bear Lake in north-west Canada. Here, 525-million-year-old animals have been exceptionally well-preserved with fully resolvable structures as narrow as 100 nanometres, or one ten-thousandth of a millimetre (less than the wavelength of light). Among the fauna known from Mount Cap is a species of *Wiwaxia*. *Wiwaxia* was a primitive form of bristle worm where the 'bristles' were modified to become protective spines and scales. Its body was short and fat and its overall appearance was that of an armoured mouse. The Burgess Shale also contains the fossilised remains of a *Wiwaxia*, although a different species to the Mount Cap type. This illustrates the importance of the Mount Cap fossils. Although they don't represent a diverse community, they contain very close relatives of the Burgess Shale fossils but are some ten million years older. This type of evidence can be used to set the date of the Cambrian explosion more precisely. *Wiwaxia* is also known from a similar period in the Spence Shale of Utah, USA. In fact the Mount Cap and Burgess Shale fossils now appear to belong to a broadly continuous belt of comparable early and middle Cambrian fossil assemblages extending from southern California through to northern Greenland and Pennsylvania.

Another extensive collection of Cambrian fossils is known from Chengjiang in Yunnan province, south China. The Chinese palaeontologist Hou Xianguang found the first Chengjiang specimen, an unusual trilobite, as a student in 1984. Hou, and his senior at that time, Chen Junguan, dedicated their subsequent years to the Chengjiang site. Extremely well-preserved specimens from several animal phyla were unearthed, and efforts to find further phyla continue at considerable pace today. A study of the ecology of the Chengjiang fauna is receiving the kind of attention previously reserved only for the Burgess Shale fossils. The Chengjiang palaeontologists' ammunition has been the age of these fossils – 525 million years old – in addition to the wide diversity of animals represented. So the Chengjiang fauna predate the Burgess fauna by ten million years and reveal that a community structure similar to that we know of 515 million years ago was already in place 525 million years ago.

Cambrian fossil assemblages that can match the Burgess Shale in terms of preservation have been, and no doubt will continue to be, discovered. But the fossils of the Burgess Shale will remain a landmark in the study of evolution, not least because of their services to the Cambrian in the macrocosm of popular science. This may seem a trivial point compared with the wealth of pure science derived from these fossils, but in the modern world of science politics are as important as knowledge, and, without the Burgess Shale's rise to fame, the expeditions which led to findings of further Cambrian fossil sites might never have been funded or been made enticing.

The $64 million question

We now know of wonderfully preserved communities of animals where a diversity of animal phyla are represented from the Cambrian, but not before the Cambrian. As stated previously, the internal body plans of animal phyla evolved some 120 to more than 500 million years earlier (depending on who you believe). Hence, the variety of internal body plans found in animals today really was once hidden within the bodies of worms, for tens of millions of years. Now we can really understand *what* the Cambrian explosion is. It is the sudden acquisition, 543–538 million years ago, of hard external parts by all the animal phyla found today (except the sponges, comb jellies and cnidarians). It is the simultaneous transition from the prototype worm-shaped or soft-bodied form to complex, characteristic shapes (also known as 'phenotypes') within each phylum, and it happened in a blink of an eye on the geological timescale. The *what* of the Cambrian explosion is now understood.

For some reason the early members of each animal phylum did not acquire their hard parts, and hence their characteristic external parts, until the Cambrian. This poses a different question – the *why* of the Cambrian explosion. Why did it take place? The evolution of hard, external parts was not a chance occurrence. It took place simultaneously in all phyla, after a considerable period during which nothing happened. This extensive correlation must have been forced by an external factor. But what factor? What caused the Cambrian explosion – *why did it happen*? This is the problem we are left with, and the aim of this book is to solve it.

Why did the characteristic external parts of animal phyla not evolve when the genetic identities were laid down in the Precambrian? Perhaps they simply did not need to. The development of complex, hard external parts from an embryo requires more energy than a simple sausage-shaped sac – why spend more energy than is necessary? And for some 120 million years or so they did not make the leap to external part development. The factor, then, that caused this leap, and made the expenditure of additional energy necessary, must have been monumental. In this book I aim to reveal the identity of this factor and hence the cause of the Cambrian explosion – the reason why it happened.

The answers proposed

A number of explanations for the *why* of the Cambrian explosion have been put forward. Unfortunately, there is strong evidence against all of them: none can stand up to scientific scrutiny. The simplistic explanation is that the general environmental conditions were uniquely befitting for evolution during the Cambrian. That is, this was simply a nice time and place for animals to evolve. This includes both the physical (non-living) and biological (living) factors within the environment. But recent finds of embryos of nonskeletised animals from the Cambrian have provided evidence against this rather circular argument. The eggs of two Cambrian animals, a jellyfish and a bristle worm, are large compared with those of their living ancestors. The considerable elbowroom within the egg, and the close resemblance of late embryos to their adult forms, are clues that Cambrian embryos hatched fully equipped to depart into the environment rather than passing through a series of less-than-proficient juvenile stages. This strategy, known as direct development, is common under harsh or unpredictable environmental conditions today. It ensures that offspring will survive rough times. For example, crabs usually hatch from their eggs as slow moving planktonic forms that drift around in the water. These young forms are easy prey for many fish and when times are hard even these meagre morsels become fish food. But if the young crabs hatch so that they can live on the sea floor, and possess colour pigments and shapes that blend into their backgrounds, they may escape the attention of predators and survive to become adults. This is not the usual method of development because a highly developed hatchling comes with a high energy cost to its parent. Direct development in the Cambrian is perhaps a surprise because it indicates that this period was not so hospitable after all. Out goes the 'nice conditions' hypothesis.

Some other explanations of the cause of the Cambrian explosion have been victims of a general misunderstanding of *what* the Cambrian explosion really is. Many scientists have launched their research to expose the *why* armed with a very misleading explanation of this event – simply, the spontaneous evolution of all animal phyla. This is *not* a fair summary of the Cambrian explosion, and one which I will name the 'misleading' explanation. Now we know that the Cambrian explosion was the spontaneous evolution of *external body parts* in all phyla, where the internal body plans of all phyla are already in place. To be fair, scientists in the past have misunderstood the Cambrian explosion through no fault of their own – the genetic evidence that tells the story of internal body plans is a recent finding.

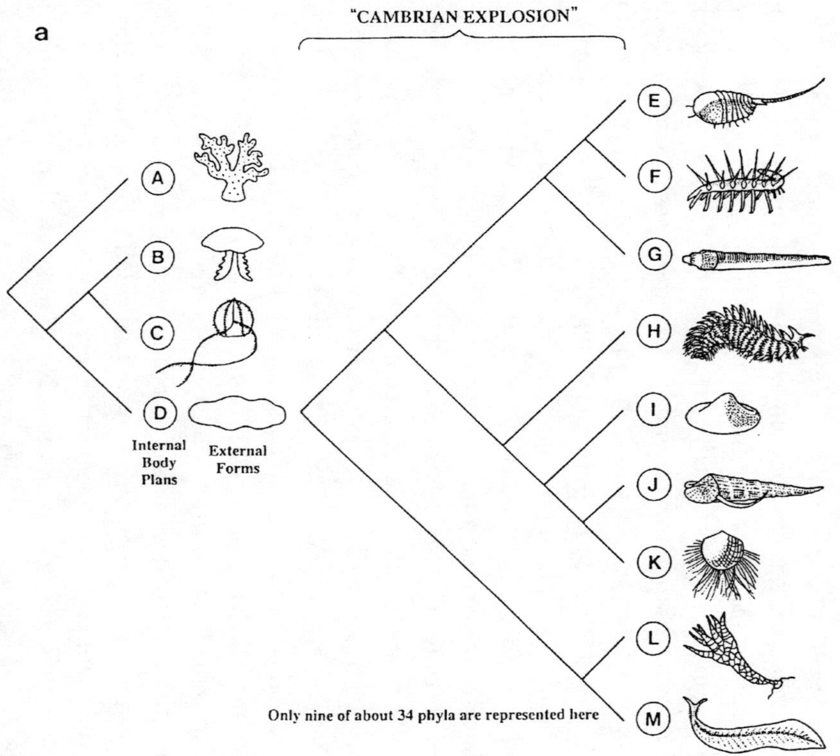

a

"CAMBRIAN EXPLOSION"

A

B

C

D

Internal Body Plans

External Forms

E

F

G

H

I

J

K

L

M

Only nine of about 34 phyla are represented here

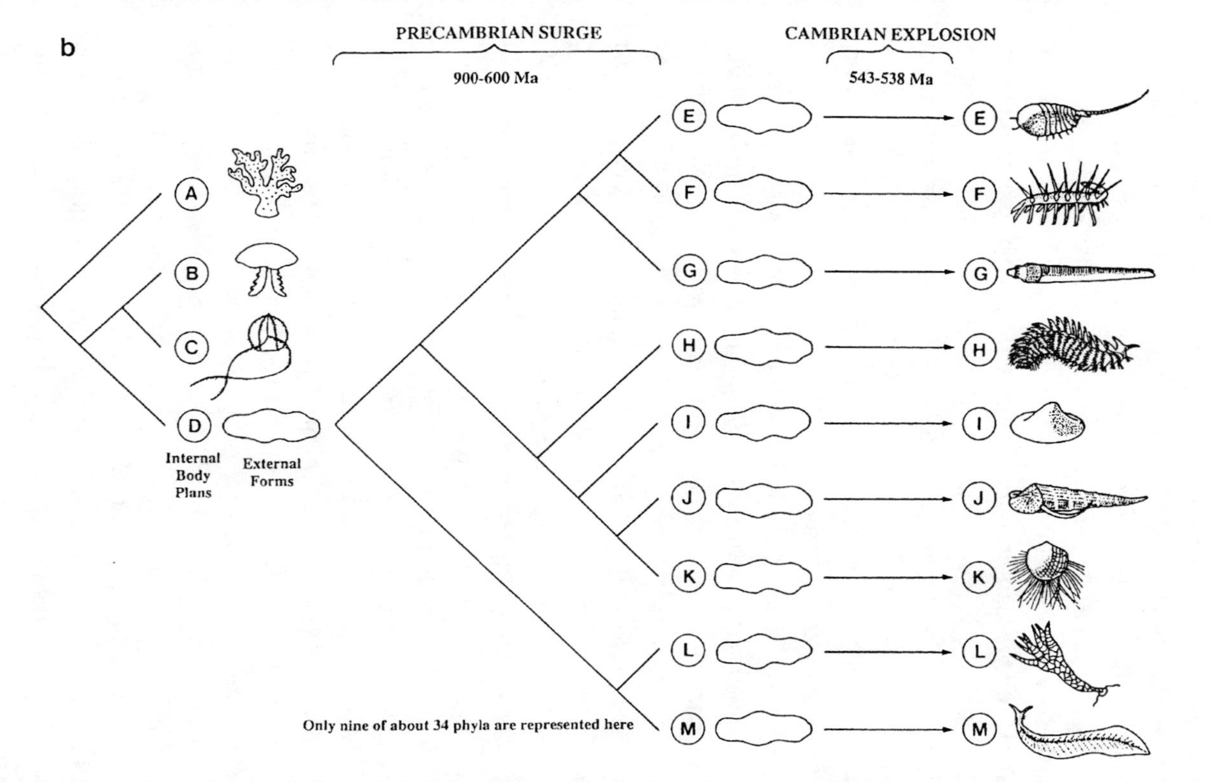

Figure 1.8 Two versions of the history of the animal phyla. From the first soft-bodied form, evolutionary branching is equivalent in both models. (A) indicates that both internal body plans and external parts diversified throughout this branching, and most theories on the cause of the Cambrian explosion have been based on this model. (B) is the correct model and properly identifies the Cambrian explosion – that it was the *simultaneous evolution* of external forms in all phyla.

Current evidence suggests that the Precambrian 'event' – the evolution of internal body plans – was not explosive but gradual, lasting tens or hundreds of millions of years. This is likely because the Precambrian 'event' concerned one animal form evolving from a previous form, and so on – a condition not affiliated with *the* Cambrian explosion. The Precambrian 'event' was more a surge in evolution than an explosion. It is possible for *the* Cambrian explosion to happen at one moment in time, but not so the Precambrian 'event'. To summarise, the old interpretation of the 'Cambrian explosion' is actually the combination of *the* Cambrian explosion and the Precambrian 'surge'. In general, the surge was the major genetic event and the explosion was rather more driven by some external factor.

The next proposals for the *why* of the Cambrian explosion suggested that the physical environmental conditions changed at the end of the Precambrian. We have already learnt that it was not a change in the environment as a whole (physical and biological factors) that caused the Cambrian explosion, but some other explanations centre on just one part of the environment. One explanation is based on a rise in atmospheric oxygen to a critical level, another on a decrease in atmospheric carbon dioxide. Oxygen and carbon dioxide are factors affecting the breathing and circulatory systems of animals. These systems are part of the internal body plans in most animal phyla, and generally affect external parts in a minor way. So oxygen and carbon dioxide levels could not have played a part in the Cambrian explosion; maybe they were involved in the evolution of internal body plans. Also, there is geological evidence indicating that oxygen levels peaked at various times before the Cambrian. Some of this evidence comes from cosmic spherules – small rocks that landed on Earth from outer space throughout geological history. Cosmic spherules contain chemicals that react with oxygen, and the degree of reactivity indicates the level of oxygen present in the Earth's atmosphere at the time of landing. And they reveal a series of peaks in oxygen levels, before, during and after the Cambrian.

Staying with the chemical theme, an additional physical environmental factor that may have changed during the Cambrian is the availability of phosphorus. Phosphorus facilitates the development of calcium phosphate skeletons, and an increase in phosphorus levels could have led to an increased development of hard external parts. But it is not just calcium phosphate that makes up the external parts of animals – other chemicals are involved too. The phosphorus argument takes no account of these. And like oxygen, there is evidence that phosphorus levels peaked before *and* during the Cambrian.

Another physical environmental suggestion for the cause of the Cambrian explosion is that continental shelf areas ('shallow' water habitats) increased at the beginning of the Cambrian. This condition may have been forged as seawater encroached on the land masses worldwide. But even if continental shelf areas increased, they were present to some degree long before the Cambrian explosion. Hence this event doesn't add anything new to the system, just more of the same.

The most recent physical environmental bid for an explanation of the *why* of the Cambrian explosion is linked to the 'Snowball Earth' hypothesis. It is thought that before the Cambrian there were spells when the Earth looked like a giant snowball. In some Precambrian times, the sun was probably some 6 per cent fainter than it is today. The consequent drop in both temperature and concentrations of carbon dioxide in the atmosphere allowed polar ice caps to grow. Ice reflects sunlight and the infra-red radiation from the sun that heats the Earth's surface. So the more ice that formed, the cooler the planet became and the greater was the potential for further ice to form. A hardline view of this idea is that all the Earth's oceans eventually froze to a depth of about 1 kilometre. A softer view is one of greatly extended polar ice caps, leaving open water to circulate around the equator. In either case, normal conditions would resume after volcanic events filled the Earth's atmosphere with enough carbon dioxide to kick-start the greenhouse effect, or global warming, causing the ice to melt. These Snowball Earth events may have taken place regularly some 2,000 million years ago, but did so at least twice during the late Precambrian, between 850 and 590 million years ago. Inevitably, for an event taking place near to the Cambrian, Snowball Earth has been nominated as the cause of the Cambrian explosion. One problem is that the scientific jury has yet to decide on the hard or soft view of Snowball Earth. The soft version provides no explanation for the Cambrian explosion for the same reason as the sudden increase in continental shelf hypothesis – namely there remained 'normal' watery environments available to host evolution. But even the hard view is open to criticism as a cause of the Cambrian explosion.

Firstly, this idea has a teleological foundation. It assumes the course of evolution was predetermined from the beginning. We are given a situation where the Precambrian worm-like bodies of all animal phyla are just itching to take on their Cambrian forms, but ice puts everything on hold. Then, when the ice has gone, it is time for evolution again. This is not an objective view. As we have considered before, why should a convenient worm shape *have* to change? If the course of evolution was predetermined, why did it not continue in the water under the ice? The second major doubt cast over this laboured explanation for the *why* of the Cambrian explosion is that the figures simply do not balance. The Cambrian explosion took place between 543 and 538 million years ago. The last Snowball Earth event ended 575 million years ago at the latest. So there is a difference of at least 32 million years between these two events. This is fact. So a Precambrian Snowball Earth event cannot explain the *Cambrian* explosion, although it could have played a role in the Precambrian 'surge'.

We are trying to explain an *explosion* in diversification, or a macroevolutionary event. In terms of external parts, changes in physical environmental conditions lead only to micro-evolution, or gradual transitions. To explain the cause of an *explosion* we need a factor that is a matter of life and death. Such a factor must be part of the biological environment – a change took place in the animals themselves. And biological environmental explanations for the cause of the Cambrian explosion have also been proposed.

One biological explanation is that collagen was universally acquired in animals during the Cambrian. Unfortunately, this only works for the misleading Cambrian explosion where collagen could have evolved in one animal phylum that was quickly to become the ancestor of all other phyla in the Cambrian. Since we know that evolution did not happen in this way, the independent evolution of collagen in all phyla would have to have happened simultaneously if this explanation is correct. The chances of this happening are extremely slim. Also, like phosphorous, collagen is not the only material used to build the hard external parts of animals.

The American biologist James Valentine, from the University of California, Berkeley, proposed that major diversification can only take place when there is an unoccupied niche (a 'way of life') to evolve into. This implies that the *why* of the Cambrian explosion is the sudden availability of niches in the Cambrian. Unfortunately, this explanation is a victim of the misleading version of the Cambrian explosion. We are not looking for an explanation of why four animal phyla suddenly evolved to become thirty-eight phyla; rather why thirty-eight phyla with different internal body plans only suddenly became thirty-eight phyla with different internal body plans *and* different external body forms. For some 120 million years this transition did not take place, yet all that time there were certainly new niches to evolve into. For example, one potential niche included a predatory lifestyle. The worm-like forms of this 120-million-year interval were basically slow-moving chunks of protein. But no animal filled this predatory niche, which may have exacted a body with hard, biting jaws and strong, grasping limbs. Numerous examples exist of the potential niches available prior to the Cambrian explosion, but for some reason these niches did not become filled until the beginning of the Cambrian – they remained potential niches. The consideration of niches is surely important, but it is not the basic explanation we are looking for. We are looking for a factor that drove all phyla to occupy all potential niches at one point in geological time. Something *very* unusual must have happened at the beginning of the Cambrian.

There may have been an increased availability of the free-swimming plants that lived as plankton in the Cambrian. This could have resulted from a major event in oceanic upwelling, which itself has been assigned several explanations. These plants were in turn a selection pressure for animals to evolve swimming limbs, so that the plants in the water could be reached, and to evolve specialised mouthparts to eat them. A short, fat worm with big lips and no teeth could never catch, let alone chew, some of the fleet-footed plants of the Cambrian. But this explanation focuses on the generation of just one new niche, not all of the niches occupied by the Burgess Shale animals. There are more than just swimming forms represented in the Burgess Shale community, so this explanation alone is not the *why* of the Cambrian explosion, but we may now be on the right track.

One of the most plausible explanations of the cause of the Cambrian explosion suggested so far was reworked recently by Mark McMenamin and Dianna Schulte McMenamin of Mount Holyoke College in South Hadley, Massachusetts. Here all feeding modes, including predation, were considered as one major factor. On the one hand, McMenamin employed modern ecological methods to resurrect a century-old idea that animals developed shells as shields against predators. But at the same time he conceptualised the entire Cambrian community in terms of a food web, where every species has its own predators and food. This conceptualisation, nonetheless, has been criticised as being simplistic and anthropomorphic.

Despite, or even because of, all the explanations proposed, biologists and palaeontologists generally are not convinced that we understand the real reason for what was arguably the most dramatic event in the history of life on Earth. Jan Bergström of the Natural History Museum in Stockholm stated in 1993: 'Why was there a radiation in the Cambrian? Our most sincere answer is that we do not know.' Four years later, Doug Erwin of the Smithsonian Institution confirmed that 'the trigger of the Cambrian explosion is still uncertain'. With this book I aim to put an end to the uncertainty and the speculation about the cause of the Cambrian explosion. I will agree with Balavoine, Adoutte and Knoll, who independently inform us that the explanation lies in a sudden change in the ecology and behavioural system of multicelled animals. But I will be much more specific.

Preview

Often in science, learning that a theory is wrong can be almost as useful as knowing it is right. The wrong answers for the *what* and the *why* of the Cambrian explosion have gradually led us to understand where to look for the correct answers to both questions. They are themselves pieces of the complete jigsaw puzzle, albeit on the edge of the picture. Having explained the correct answer to the *what* of the Cambrian explosion in this chapter, I will set out in the remainder of this book to present my new explanation of the *why* of the Cambrian explosion, something that has become known as the 'Light Switch' theory. To uncover the real cause of the Cambrian explosion we need to put together *all* the pieces of the puzzle. The next seven chapters of this book will be given over to the more significant pieces. In the course of these chapters I will construct a multidimensional picture of how life works today, what happened during the course of evolution on Earth and, consequently, how life worked at different times in the past.

The following chapters will bring together the most unlikely of subjects, from ancient churches to impressionist paintings. At the turn of the twentieth century, the president of the Inventors' Association resigned his position after claiming, 'Everything that could be invented has been invented.' He was not missed. This book will demonstrate the rewards of exploring laterally and how science can benefit from an interdisciplinary approach.

In the next chapter I will examine fossils in more detail, providing examples of the information they have yielded in the past. But an explanation of the Cambrian explosion needs more than palaeontological evidence; it needs biological evidence too. As many clues can be found from studying living ecosystems as can be found in the fossils of the Cambrian animals themselves. The solution I propose draws on clues from all over science. By moving through time to the living world and on to my own specialist subjects, I will explain, in Chapter 3, how modern animals appear coloured or invisible. I will demonstrate the sophistication of the colour-producing systems of today's animals, something we know very little about in extinct animals. A central theme will be that light is the most important stimulus to animal behaviour in the vast majority of today's environments – those exposed to light.

The case for light as a major stimulus today will be strengthened in Chapter 4 by examining the other side of the story – life in the dark, in caves and in the deep sea. Here, the importance of light will become even more apparent, not just in animal behaviour but also in evolution. In Chapter 5 I will compare the rates of evolution in two groups of seed-shrimps which began their histories in different environments. One group lives in the open sea, the other in marine caves. By taking a closer look at the group from the open sea, it will emerge that light has driven their evolution, while those in the dark have barely changed from their primitive ancestors. The result is that the open-sea seed-shrimps are considerably more diverse than they are in dark caves. The role that light can play in evolution will also be demonstrated using marine isopod crustaceans (to which woodlice, or slaters, belong), where we will join Jim Lowry's SEAS expedition in the Pacific Ocean, and also using crabs and flies.

In Chapter 6 I will lighten the mood a little with an exploration for colour in ancient, extinct animals. Bones and other hard parts that may become fossils are physical structures. Some colours today result from physical structures, albeit microscopic. Could such micro-structures also preserve in the fossil record? Potential will be unearthed in fifty-million-year-old beetles and 180-million-year-old ammonites. Then the pages of the history book will be turned back even further ... If the original colour alone of an Egyptian statue can tell us that it once housed the Book of the Dead, just think how much can be learnt from finding colour in fossils.

To balance the information provided on colour in animals, Chapter 7 will introduce the variety of eyes. It will show that all animals have to be adapted to the existence of eyes not only in terms of their colour, but also in their shape and behaviour – all factors affecting an animal's appearance on a retina. When this retina belongs to a predator, the image formed on it becomes a matter of life and death for the potential prey. But is the danger of visual appearance a recent one? The history of predation will be discussed in Chapter 8. By returning to the fossil record I will show that eyes, predators and probably the link between them go back a long way. But exactly how long? This will become a fundamental question.

By the beginning of the penultimate chapter the reader will have all the clues necessary to decipher the probable cause of the Cambrian explosion. In many ways it seems the most obvious explanation, but to reach it one must take this indirect, winding road. Encountered along the way will be a number of unfamiliar but fascinating examples of the sophisticated and finely balanced ecosystems that exist in nature – but to begin with, it's back to the bare bones and a modern perspective on the lifeless rocks once kept safely within dusty Victorian display cases. Now we are bringing the past to life.

2

The Virtual Life of Fossils

Nothing ever becomes real until it is experienced

JOHN KEATS

Beginning, as it were, with the very beginning, Chapter 1 summarised a history of life on Earth. In this chapter, the evidence used to create such a story will be examined, making a closer inspection of the rocks. But here time shall be traversed from today, travelling back to the Cambrian via some landmark attractions. And good old-fashioned fossils will provide the attractions.

Although the study of evolution is increasingly becoming consumed by genetic studies, the inferences from genetics are, and always will be, theoretical. The genes of many living species have been exposed, but the animals we see today did not evolve directly from each other. Intermediate stages were involved – species, for instance, that became extinct. So in order to reveal evolution, the genetics of the living *and* the extinct are required. And of course the extinct genes are, barring a few exceptions, subject to theoretical fabrication.

Fossils, on the other hand, are factual. They are literally hard facts that we cannot ignore. Around a decade ago, molecular sequences pointed to a Cambrian explosion that occurred way back in the Precambrian. The fossil record, which places an Early Cambrian label on the grand event, was thus contradictory and appeared to be standing in the way of progress. But palaeontologists stood firm, reminding us that fossils were not optical illusions. When 350-million-year-old rocks are split to reveal the fine details of a bony fish, then bony fish *did* swim in Earth's waters 350 million years ago. When rocks formed under similar conditions, but from 550 million years ago, are consistently found without bony fish, eventually we must conclude that bony fish did not exist during this time. However, it would be equally foolish to ignore the genetic evidence, and indeed by reconciling the fossils and the genes a true picture of the Cambrian explosion has been painted. But whichever way they are looked upon, fossils are precious to the study of evolution. And they certainly justify a chapter of their own in this book, where the subject of fossils will surface again during discussion of seemingly unrelated topics.

It was the role of fossils in revealing the paths taken by evolution which contributed heavily to the previous chapter. The main purposes of this chapter are to expose the tricks used in creating this knowledge, but also to demonstrate that fossils have much more to say. The history book, 'The History of Life', conceptualised here contains two-dimensional pages. The next task is to pump blood into the flattened veins of fossils and let them spring from the pages, so ancient animals can be seen doing what they once did. The application of engineering, physics, chemistry and biology can indeed transform a load of old bones into a virtual 3D world, perhaps millions of years old, where animals run, fly, gallop, burrow, eat and avoid being eaten.

Fossils can add some surprising details to the past, and they will provide considerable hard evidence towards the Cambrian enigma that this book attempts to solve. The individual cases in this chapter will provide a flavour of palaeontology in the twenty-first century, and constitute tools for the evolutionary trade. The art of Sherlock Holmes and modern forensic science will be reconciled with that of dinosaur specialists and religious artists. Fossil leaves will be employed to aid the palaeo-meteorologist. The technology of car designers will bring 400-million-year-old 'worms' and arthropods back to virtual life on the computer screen. And the biology of living organisms and principles of Scuba diving will help to solve the 'ammonite mysteries'. But to begin I will ask the question: 'What, exactly, is a fossil?' The answer to this is not so obvious, especially when the remains of some extinct species are so 'fresh' they can literally be brought back to life.

The youngest fossils

I have a colossal, antiquated book on the fauna of Earth. It is entitled *Knight's Pictorial Museum of Animated Nature* and is now in its seventh generation within my family. Between the heavy, morbid black covers exist brief descriptions, biological data and woodcut illustrations for thousands of species. Some of the illustrations are quite primitive, especially the unnatural poses of monkeys quite clearly based on stuffed museum specimens. The kangaroo drawings appear like those made by the first Europeans to reach Australia, and the story is similar for the

Figure 2.1 Butterworth's 1920s illustration of *Diplodocus* walking, crocodile-style.

American buffalo. A quick glimpse of a very unfamiliar form can result in a reconstruction with a more familiar form in mind. A buffalo could become cow-like, and a kangaroo could acquire some of the features of a hare. Here lies a lesson in fossil reconstructions – extrapolation can be risky, at least beyond a reasonable point. Crocodiles may be the closest living relative to certain dinosaurs. Although it may be safe to infer a similar scale-like skin texture, as we can confirm from recent finds of fossil skin, the sluggish quadrupedal form with a belly that scrapes the ground is probably a characteristic of the crocodile only. Yet pioneers of dinosaur reconstructions depicted the *Diplodocus* with its belly scraping the ground. That's fine – we need mistakes from which to learn (and mistakes are everywhere in science). Nowhere is this principle of extrapolation more dangerous than in the colour of extinct animals, as will be demonstrated later in this book.

Knight's Pictorial Museum also contains information on fossils. At the interface of the living and fossil species lies the dodo, an animal we know so much about through the written accounts of seventeenth-century travellers who descended on its native Mauritius, yet it has been extinct since at least the time of *Knight's Pictorial Museum*. But an even more detailed account of behaviour is given for the great auk and Tasmanian tiger, both of which, distressingly, appear in the section of living animals. The great auk and Tasmanian tiger are now extinct.

The feet and the skin from the beak of a dodo are preserved in natural history museums in London and Oxford. A great auk in its entirety can be seen stuffed in a penguin-like pose in London, and complete Tasmanian tiger specimens, which survived to see the twentieth century, are more common. Maybe there are many more cases like these. We are living in the harshest extinction event of all, which highlights the growing importance of natural history museum collections. One day I became interested in the colour of stick insects, and while I was exploring the entomological cabinets of the Australian Museum in Sydney, my attention was drawn to a giant specimen from Lord Howe Island in the Pacific. Unfortunately my request for a loan of this fragile specimen was rejected on the grounds that it was collected over a hundred years ago and was the last of its kind. But can we classify specimens that contain their original, organic parts as fossils, even though their species are now extinct? Maybe the age (relatively youthful) of the specimens (in geological terms) in these particular cases provides a strong bias against a fossil categorisation – our not-too-distant relatives could have collected them.

The question as to what defines a fossil becomes more interesting when the subject derives from a more distant epoch. The first mammoth appeared 150,000 years ago, into the second to last Ice Age. The mammoth spread through northern Asia, America and Europe, sharing its environment with giant ground sloths, sabre-toothed cats and big-horned bison. Precisely 20,380 years ago, one individual, 8-ton, 11-foot-tall male mammoth died on the frozen plains of Siberia at the age of forty-seven, thirteen years short of the average life span of a mammoth. If ancient animals are considered in this way, they become animals that once lived, rather than animals that are now extinct. To effectively bring an extinct animal back to life is a palaeontological goal, but in the case of the mammoth we have evidence well beyond the norm.

Much is known of the mammoth's lifestyle through discoveries of ancient human cave dwellings. Piles of bones and tusks belonging to mammoths have been found alongside stone-pointed spears, suggesting that humans were mammoth hunters. And they were probably significant mammoth hunters. Numerous Ice Age caves have been discovered with pigments preserved – primitive paintings depicting scenes of large-scale mammoth hunts. These pictures have even prompted theories that the mammoth was the first species to be wiped out by humans. Maybe if we had earlier had preserved specimens of the mammoth we could have conducted forensic examinations to discover the extent of hunting with spears. Well, now we have one.

One day in 1997, a nine-year-old Russian boy from the Zharkov family set out to hunt reindeer in the frozen wastes of Siberia. All seemed quite normal until an unusual whitish object came into view against the blue horizon. That object became a pair of objects as the boy approached, and soon they could be identified with accuracy. Protruding from the frozen ground, or permafrost, were the tusks of a mammoth – the individual that, it would transpire, had died 20,380 years earlier. Such a sight was familiar to the rest of the Zharkov family – mammoth tusks no longer make the news in Siberia – but there was something different about this particular find. The Zharkovs brought these tusks to the attention of the scientific world because they were attached to a block of ice with signs of flesh and thick tufts of fur. That made scientists sit up and listen. The first country to secure funding for a mammoth autopsy was France.

Two years later, in 1999, an unusual operation by French Arctic scientists began. A Russian helicopter was employed to raise a huge, cubic block of permafrost, complete with massive tusks projecting. Within this block, it so happened, was a complete mammoth, almost perfectly preserved in its icy tomb. The mammoth, initially frozen to –50°C by searing winds, was airlifted 200 miles to the city of Khatanga. If the sight of that alone did not create some amusement, the event that followed certainly did – for six weeks scientists stood around the mammoth defrosting it with hairdryers. But the scientific team had the last laugh when they became the owners of a museum-quality mammoth specimen, complete with DNA. The hairdryers not only warmed the ice but also dried the skin and muscles, thus aiding preservation.

At the moment, the French team is conducting a thorough forensic examination on the 20,380-year-old mammoth carcass with the aim of determining the cause of death. This could provide support for the theory of a human-driven extinction, or supply evidence towards an alternative idea that the species succumbed to malnutrition following a dramatic climate change. A spear would leave its telltale impression in frozen flesh, but maybe not so in a skeleton. A skeleton also would provide little evidence of malnutrition. So there are certainly limitations to interpreting the past using only the bones, but, as we will see, we have further tricks up our palaeontological sleeve.

As to whether this mammoth could be considered a fossil, the answer is really not so important. Here the original organic material is preserved, like the skin and bones of Egyptian mummies. In the true tradition of a fossil, a carcass is entombed within a material of some description before decomposition by microbes takes over. This can happen via sedimentation, when mineral particles falling out of the water blanket the carcass on their way to forming the sediment or substrate that constitutes the sea floor. Then minerals from within the substrate replace the organic material. The precise forms of the 'replacement' minerals become different from those in the substrate; thus the fossil is separated and easily identified from the surrounding matrix. But sometimes only part of a newly deceased carcass becomes fossilised and the remaining organic material is preserved unaltered. Since this balance can shift in either direction, it is academic whether we apply the term *fossil* to an ancient specimen with 1 per cent replacement minerals and 99 per cent organic material, but not to a specimen with 100 per cent organic preservation. Either way the dead animal has left its mark for palaeontologists to find.

Additionally the fossilisation process itself can occur in varying degrees of complexity. The outlines of bones only can be saved as fossils. But then sometimes the skin, organs and internal parts of bones can be entered into the fossil record too. When the fine detail is preserved, physical information can be extracted equally from a true fossil or preserved organic material. We know that mammoth tusks were optimally strong due to their construction. The stacks of thin, corrugated layers of alternating material provide greater strength and toughness than do either thick layers of alternating materials or stacks of thin layers that are flat in profile. Plywood and corrugated iron are strong for these reasons, respectively. This information can be extracted from both truly fossilised remains and original organic specimens. Less well-preserved fossils, on the other hand, bear only the outline of tusks, providing information on their size and shape only. But there is one important difference between the well-preserved fossils and organic remains, and one that is showing signs of great scientific potential – the preservation of nucleic acids.

The boundary between organic remains and classical fossils becomes increasingly fuzzy when the organic subjects are seventy million years old. Surely remains this old must be considered fossils? Insects that lived seventy million years ago have been preserved in amber, in all their organic glory. Flies coming to rest on tree trunks today occasionally find themselves sticking to the yellowish sap that seeps through bark. The more the flies struggle, the further into the sap they sink, quicksand-style. Seventy million years ago, flies came to a similarly sticky end. The sap eventually hardened and entombed the flies for ever.

This hardened sap, called amber, provides a barrier to microbes and chemicals, so the organic material of the fly remains unchanged. Embarrassed by the age of these specimens, palaeontologists have coined the term 'sub-fossil' for such nonconformists. The flies in amber, particularly the blood-sucking mosquitoes, are also the most famously controversial group to be considered for the preservation of nucleic acids.

Some of the more spectacular dinosaurs lived seventy million years ago and were probably the victims of mosquitoes. The idea that dinosaurs could be brought back to life based on dinosaur blood preserved within ancient mosquitoes is now a distant one. What rained on this particular parade was contamination – the apparently ancient DNA from dinosaurs was in fact recent DNA, from a contaminant within the molecular lab. Now it is generally believed that nucleic acids cannot survive such periods of millions of years. But the methods planned for converting genomes into a living, breathing *T. rex* have been retained.

Microbiologists routinely revive 10,000-year-old microbes from Antarctica. Living microbial spores can be dispersed by the wind, and some have the misfortune to land on the ice of Antarctica, where the spore cells immediately become dormant. They shrink in size and shut down all metabolic activity.

The Russian station at Vostok is one of the most uninhabitable in Antarctica, situated at the very centre of the continent. At Vostok the ice is drilled and cores are removed, and the ice at the bottom of a core can be up to 500,000 years old. In 1988 an American microbiologist found spores locked in part of a core containing ice 200,000 years old. Miraculously, on warming the spores live bacteria emerged, which could be cultured as if 200,000 years had never elapsed. This signals hope of reviving other nucleic acids up to a similar age, and is viewed as an unconditional green light by the mammoth team.

The French owners of the 20,380-year-old mammoth have high hopes of extracting DNA from the frozen cells and cloning a new mammoth – one that walks and does the things a mammoth did, things we would like to know. One Japanese scientist likes the idea so much that he is scouring Siberia and Alaska for a frozen mammoth of his own.

Similarly, nucleic acid from an appropriately preserved Tasmanian tiger ('thylacine') pup has been extracted, with cloning intentions, at the Australian Museum. Here, cloning methodology, which utilises the closest living relative as a surrogate mother, is under investigation. In addition to cloning such old DNA, pitfalls to consider include its compatibility with chromosomes from a different species, followed by the acceptance of a foreign embryo by a surrogate mother. Then, if the cloning is successful, scientists must aim to avoid sterile creations. Mules, for instance, are almost invariably sterile because they amass an odd number of chromosomes – thirty-one from the donkey parent plus thirty-two from the horse parent. The sixty-three chromosomes in the mule's body cells divide randomly into thirty-one or thirty-two in the reproductive cells. When two mules mate, the pairs of gametes are so unevenly matched that the chromosomes simply cannot pair up. But if novel cloning methods did succeed on the Tasmanian tiger, a new science would dawn. And in any case, sequences from ancient nucleic acid would be useful to fit into those evolutionary analyses that otherwise rely on predictions when dealing with creatures that are extinct.

Ancient DNA has been put to another use in mapping the geographical history of disease. The plant pathogen responsible for the nineteenth-century Irish potato famine, *Phytophthora infestans* (late blight), has been identified by sequencing DNA from museum and herbarium samples of infected potatoes and tomatoes. The ancestral clone of this late blight was believed to have arisen in Mexico and been widespread during the past century. But recently the strain responsible for the potato famine was found to be different, having a South American origin. Late blight remains active today, and if its true history is known, future geographical spread may be predicted. This case also provides further justification for natural history museum collections – as nucleic acid banks that preserve the *genetic* diversity. But returning to the subject of bringing ancient, extinct animals to virtual life, if the genes can't help us at present then we must return to the fossils and so-called sub-fossils.

Our supply of sub-fossils begins to dry up in rocks older than 100 million years. But exceptions do exist, and indeed provide some very important evidence in this book. For now, though, it's back to the genuine, reliable fossils and what they can teach us.

Old bones, new science

Palaeontology has been strong for well over a century. Fossils themselves have provided solid foundations, and the house of palaeontology has risen with walls of equal strength, built with successive blocks of compatible theory. Throughout this construction fossils have always been reliable, but their interpretations – the building blocks – are occasionally flawed. And misconceptions that establish themselves in palaeontological law must eventually be amended. One of these flawed building blocks could exist near the base of the mammalian evolutionary wall.

Most mammals, including ourselves, nurture their developing young within a uterus and are called placentals. Placentals were thought to have originated in the northern hemisphere more than 100 million years ago. Later they spread throughout the globe, and in doing so forced the other two groups of mammals – the egg-laying monotremes and the pouch-bearing marsupials – to retreat. The monotremes (like the platypus) and marsupials (such as kangaroos) retreated towards Australia, their main place of existence today. In fact, the first land-based placentals were believed to have migrated to Australia a mere five million years ago. A nice, neat story that became established in zoological textbooks, an evolutionary classic. But now to throw a spanner in the works – to be precise, a 115-million-year-old jawbone found by a British volunteer working on a beach in Melbourne, Australia, in 1997.

The tiny jawbone, just 16 millimetres long, holds eight of the most controversial teeth ever discovered. Three of the teeth are molars and five are premolars – a characteristic of placentals and not marsupials, which usually have four molars and three premolars. Then there is the shape of the teeth. They are adapted for slicing and crushing food, a feature not found in monotremes. Also, one premolar seems to be departing from the standard triangular form and is almost halfway to becoming the more elaborate form characteristic of molars. Again, this fits with the placental code but not with that of marsupials.

Nearing the outcome of this controversy, the jaw in question has been extrapolated on a computer screen to become a virtual shrew – an insectivorous placental named *Ausktribosphenos nyktos*, after its discoverer. Could this interpretation represent a crack in an esteemed palaeontological wall? The notion of a placental mammal running around in Australia 115 million years ago would certainly turn our view of mammalian evolution upside down. However, this story has yet to reach a satisfactory conclusion.

A new idea is emerging from the University of California at Berkeley that *A. nyktos* is neither a monotreme nor a marsupial, and not even a placental, but rather a new group that was either converging with other mammals or running parallel with them, before eventually dying out. The evidence for this theory derives from further refined analyses. It happens that the shape of the depression at the back of each molar is uncharacteristic of placentals.

So when the *A. nyktos* building block is cemented to the mammalian evolutionary wall, is the construction strengthened or does it fall? For now we must lay down our tools and search for the complete trail of mammalian clues that wait to be unearthed. Maybe some earlier finds would benefit from re-examination. Certainly, modern analytical methods in palaeontology are becoming increasingly refined and sophisticated, and can reveal a surprising wealth of information from even the tiniest portion of a skeleton. And advancements in reconstructing the extinct will become a theme of this chapter. Sometimes it will cause established pillars to fall, and sometimes to rise even higher. But first it is worth pursuing the idea of ancient animals migrating between continents, and how this can be possible considering the immense perimeters of water we see on today's globe.

The active Earth

The landmasses that form our continents are not static; they are plates of rock that are continuously moving around within the Earth's crust. These movements are not just horizontal, they are vertical, too. They affect the land both above and below the water. Geological faults are evident at the deepest regions of the ocean, where plates moving in opposite directions eventually tear apart, presenting an opportunity for molten lava from beneath to seep through the gaps formed and into the water. This can be the making of a hydrothermal vent – the black smoker introduced in the previous chapter. On a grander scale, the Hawaiian islands were formed by this very activity. Comparable faults on land can result in the eruption of volcanoes, and the tearing apart at one part of the globe means a crashing together at another part. The Himalayas formed when the Indian plate, once bordered only by sea, crashed into the Asian landmass. This event saw the pre-existing Asian coastline forced upwards, effectively turning the Earth's plates upside down.

Deserts have not always been areas of desolation. Coral reefs have not always existed where they are found today. In the scheme of global history deserts and coral reefs can be associated by their geography – they may share the same geographic coordinates. The Great Basin area of Nevada, Utah and California forms one of the most significant deserts in the United States. Yet in the rocks of higher altitudes can be found fossilised ecosystems that existed some 510 million years ago – underwater. Corals, moss animals, arthropods and many other animal phyla are represented in abundance after their otherwise shallow graves were entombed forever by a fateful wave of mud. The mud turned to stone and fixed the life forms forever, but in different geographic locations through time. Their graves were transported with the movement of the Earth's plates. First they were lifted out of the water, then high into the air, until they are exposed on a mountainside today. In fact the Burgess Shale fossils embarked on a similar journey. But the Earth's plates continue to move, and maybe in another 100 million years these fossils will travel full circle and return to their watery origins.

Reconstructing ancient environments

The conception of a mobile Earth's crust is known as plate tectonics. This subject becomes extremely important when contemplating the original environment of a fossilised animal. A point on the Earth's crust can change its longitude and latitude, and it is perhaps the latitudinal change that is the more significant. That is the movement most responsible for a climate change. So a fossil found in a hot desert need not belong to a hot desert dweller. But the biggest clues to a fossilised animal's precise environment can be found in the surrounding fossils, particularly the plants. Aquatic plants are quite distinct from their counterparts on land. But Cambrian life was exclusively marine, so in order to improve on Cambrian biology we should distinguish further than between simply land and aquatic environments. And if one looks even more closely at the entire fossil community, one really can be more specific. The presence of photosynthetic algae in a fossil assemblage indicates that this community lived under reasonable levels of sunlight, placing the extinct environment within the photic zone – between the ocean surface and around 90 metres in depth. Similarly, biology can be inferred from fossilised land-based organisms. For instance, we are beginning to map the fine variations in the external skeletons, or exoskeleton, of living beetles.

Beetle exoskeleton is effectively constructed of thin layers, laid down parallel to the outer surface. If the individual layers are relatively thick and corrugated, then the beetle can withstand high temperatures. If there are many pores in the exoskeleton then wax can be secreted to prevent it drying out. A combination of both characteristics indicates an adaptation to deserts. Beetles from temperate climates tend to possess flat layers in their exoskeletons, where all the layers are thin except for a very thick outer layer which provides physical protection. The exoskeletons of cold-adapted and aquatic beetles are different again. So the structure of beetle exoskeleton could be considered an indicator of temperature or other properties of an environment. But can this theory be applied to the geological past? Interestingly, well-preserved fossilised beetles exist with their exoskeletons intact, such as those from the twenty-five- to thirty-million-year-old fossil site at Riversleigh in Australia. Maybe further information on the original Riversleigh environment really can be deduced from its beetles. And the potential for linking fossil anatomy with ancient environments has been bolstered by a study on plant leaves.

For over a century it has been known that increased levels of carbon dioxide in the atmosphere lead to an increase in temperatures. Similar conclusions were more recently drawn from analyses of air trapped in ice cores 420,000 years old, an age where temperatures are known. Unfortunately, studies on ice cores are restricted to the last half a million years. So to link carbon dioxide to some of the really important events in Earth's biological history, new ways of tracking the history of this gas are needed. One ingenious new way could be to use our extensive collections of fossil leaves.

Plants require carbon dioxide for photosynthesis. The gas is taken up through valve-like pores that occur on the surfaces of leaves. It is understood that the past 200 years have witnessed increased carbon dioxide levels as a result of industrial fossil fuel consumption. It is also known that plants have responded to this increase by producing fewer pores on their leaves. In fact there is a distinct inverse relationship between the concentration of carbon dioxide in the atmosphere and the density of pores on leaves. And now this relationship has been exploited by a palaeontologist in possession of fossil leaves from *Ginkgo* trees and their like, up to 300 million years old.

Within the collection rooms of the University of Oregon, dust was blown from the stacks of ancient leaves, which overlap in the fossil record, and the proportion of pores was determined. This resulted in a complete count of pores over the past 300 million years. From the pore counts, the levels of carbon dioxide have been predicted over this vast period. In turn, 300 million years of atmospheric temperature has been discerned. Impressive work, proving again that it can be worthwhile waking the sleepy, forgotten museum collections.

Marine geochemical data accurately predicts the temperature of the more recent part of geological history – and this matches the predictions from pore data. To test the predictions further back in geological time, one can turn to the record of sedimentation and the oxygen-isotope record of marine fossils. The oxygen-isotope data shows only trends in temperature over the past 300 million years, but the peaks and troughs do conform well with those from the pore data. Both data sets infer that periods of low carbon dioxide prevailed between about 296 million and 275 million years ago, between thirty million and twenty million years ago, and during the past eight million years. And the sedimentary record of glacial deposits in high-latitude regions indicates comparable trends. The periods of low carbon dioxide do appear to coincide with the periods of cool, 'ice-house' modes of Earth's climate history. But the pore data is the most useful because it has the finest resolution. This information could be invaluable when considering the cause of extinctions of ancient fauna that occurred at precise moments in geological time. Global warming or cooling could indeed push animal chemistry beyond critical barriers.

An opinion from Utrecht University in the Netherlands agrees that concentrations of carbon dioxide will emerge as the main factor of temperature and climate during the past 500 million years. But it is warned that plenty of other ingredients would have been added to the climate cauldron at different times and to different extents in the past. Changes in the configuration of continents, topography such as mountain building and ocean circulation can all have a profound influence on climate. But there are also planetary factors to consider, such as changes in the Earth's orbit or the angle of its axis, as well as solar brightness. Any of these elements could have affected atmospheric temperature and played an indirect role in major evolutionary events over the past 500 million years. But then there are suspects other than temperature which have the potential to induce macro-evolutionary events. Because the enigma central to this book lies beyond the 500 million years in question here, I can afford, fortunately, to leave this problem to others. Nevertheless, this has been a nice demonstration of how fossils can indicate past climates and ultimately help to reconstruct ancient environments. Now we can continue along the palaeontological path and consider the animal inhabitants of those environments and the marks they have left behind.

Palaeontology – the first forensic science

The word *fossil* derives from the Latin, meaning something dug up. Until the eighteenth century, any unusual object dug out of the ground was known as a fossil. In medieval Europe crystals such as amethyst and ancient man-made arrowheads were considered fossils. To North American Indians, dinosaur bones were thought to be the bones of giants that once populated the Earth. But what were they supposed to think? Today we are able to travel globally and enjoy the benefit of international knowledge. We are all familiar with tigers, elephants, emus, sharks and crocodiles even if we are unlikely to meet them all in their natural environments. But what might have been the thoughts of ancient Greek adventurers as they first set foot in Egypt, to be confronted by a crocodile? Such a creature would have been no more alien to the ancient Greeks than a dragon is to us today. Maybe Greek mythology was not so unbelievable in 500 BC. Or maybe thoughts of evolution were formulating in the minds of disparate, ancient people, as they were in those who lived nearer to the Darwinian age. Of course, any such thoughts must have been kept to the individual's own self, and the safer option of mythology formulated.

Amon was an ancient Egyptian god often represented as having the body of a man but the head of a ram. The ammonoids were a group of molluscs, long extinct, related to the octopus and squid. They possessed shells that were often coiled spirally and are found commonly as fossils today. As their nomenclature suggests, fossil ammonoid shells, or ammonites, were thought to be the horns of Amon. Admittedly, some ammonites do look like rams' horns, but ammonites can also resemble the shells of a living marine animal also related to squids – the nautilus. So can one employ the nautilus to bring the ammonoids back to virtual life? This is a significant question and, before jumping to any premature conclusions, it is worthwhile examining the techniques available to perform such a feat, which include the forensic methods for reconstructing human images. These methods have even been employed to reconstruct the most famous image of all – the face of Jesus.

Researchers have devised what is said to be the closest possible likeness of the historical Jesus, producing an image far removed from centuries of convention. The skull of a Jewish man from a first-century burial and the latest forensic techniques were combined to create a virtual image that challenges the stereotype in use in art since the Renaissance.

Before the second century Judaic tradition upheld a ban on the pictorial representation of God. Thus only the symbolic representation of Jesus could be depicted, bestowing the form of a fish or a lamb. St John's Gospel includes the statement 'I am the good shepherd', and in the earliest figurative representations of Jesus he was portrayed as the Good

Shepherd. Later, when Christianity replaced the Roman Empire, Jesus was boldly illustrated as the King of Heaven, and gained the features of the stereotypical Roman aristocrat – he appeared older, more authoritative and beardless. But the Byzantine Church always preferred the bearded Jesus, and so this image became the standard everywhere. Hence a pillar of credibility was constructed that, like those of palaeontology, has proved difficult to topple. Giotto and Raphael among others continued with the bearded tradition that has remained popular up to the present day.

In 2000, road construction workers unearthed a group of skeletons in Jerusalem. Israeli archaeologists studied the alignment of the graves and the artefacts in the surrounding earth to conclude that the burial site was first-century Jewish. All of the skulls found were quite distinct from others of different ages and of different regional tribes. One skull was selected as being a good representative of the group, and typical of the kind of person that would have lived in Jerusalem in the first century AD.

Skulls determine the shape of a face, including the eyebrows, nose and jawline. To bring the Jerusalem skull to virtual life, it was handed over to a forensic expert in England, at Manchester University. Strips of clay were layered upon a plaster cast of the skull, in the proportions known from human postmortems. (This method was employed successfully to identify the remains of a King's Cross fire victim in London in 1987, and can generally boast a 70 per cent success rate for similar identifications. In that case the head still had skin, but the skin colour, and colour and style of the hair, remained in question.)

Fossils of plant life found in Jerusalem from the time of Christ were used to back up ancient texts on climate history. The climate was resolved with precision and the model of Jesus was given dark olive skin, appropriately. This contrasts with the pale, delicate complexion of previous depictions. But still the hair and beard of Jesus remained in question, and the fashion of Jesus' times became important to their reconstructions. The only useful pigments to have been preserved were not contained in 2,000-year-old hair samples, unfortunately, but those in first- and third-century frescoes of synagogues in northern Iraq. These depicted Jesus with short curly hair and a trimmed beard, a style which would be accommodated in the new reconstruction. We have to assume the hair was dark brown.

This is not the true face of Jesus but is probably the most accurate interpretation ever created. Here, archaeological, palaeontological and anatomical science have been united to replace artistic licence. But can we employ multidisciplinary science to bring fossils to virtual life? Or specifically, can we use the living nautilus to breathe life into its extinct relatives, the ammonoids? This could really help solve another mystery – why most ammonoids became fossilised near the interface of sea and land.

For about 500 million years the cephalopod molluscs, including the octopus, squid, cuttlefish, nautilus and ammonoid, have been among the most successful of marine animals. Today squid alone are numerous enough to sustain the world's population of their major predators – sperm whales. Squid possess internal shells. The nautilus, on the other hand, has an external, coiled shell that approximates a logarithmic spiral. Its squid-like tentacles, eyes and jet-propelling siphon protrude from the open end of its shell, snail-style. The shells of the nautilus and ammonoids are similar, and are divided internally into several chambers, each separated by chamber walls. The fossil record of ammonoids is extensive, and the chamber walls tend to preserve well. So what was the function of the shell chambers?

The most obvious reason why ammonite shells have chambers is that their dividing walls provide strength for the shell. Sir Eric Denton, of the Marine Biological Association of the UK in Plymouth, conducted many famous studies on the lifestyle of the living nautilus, and came up with evidence that denied the shell-strength proposition. One study demonstrated that a nautilus shell remains fully intact as the pressure of its surroundings is increased. That is until a critical pressure is reached where, without any warning suggested by cracking, the entire shell shatters. This characteristic was linked to the natural environment of the species – it lives from shallow seas down to waters just prior to a depth at which existence would be perilous, the environment that accommodates its critical pressure. The margin of safety is slight. Examination of shell fragments indicated that shattering at critical pressure was a characteristic of the shell-wall construction, and the chamber walls did not make a difference to the overall pressure tolerance. So the living relatives of ammonoids indicated that the purpose of the shell chambers was not linked to strength.

The living part of the nautilus occupies the first and largest chamber, which is open-ended. Each chamber thereafter has an additional character – a thin tube running through its centre and through the chamber walls, terminating in the last chamber and taking on the spiral shape of the shell. Similar tubes are evident in ammonites, and in the nautilus this tube is known to contain living tissue. In terms of the body volume, the tube tissue is a minor part of the animal. But in terms of the animal's behaviour, it constitutes a major organ. The role of the tube tissue is to transport water into and out of the otherwise air-filled chambers, and so regulate buoyancy. This means that the nautilus can move vertically in the water with apparent ease.

Studies on the internal tubes of ammonites revealed similar properties to those of the nautilus – that water could have permeated the tube walls through gaps along its length. This led to a lifestyle reconstruction for the extinct ammonoids. They were portrayed as poor swimmers going forwards and backwards, but highly adapted for moving up and down. And then further gaps in our biological knowledge were filled – the image of an ammonoid moving vertically in the water column at speed was linked to its food.

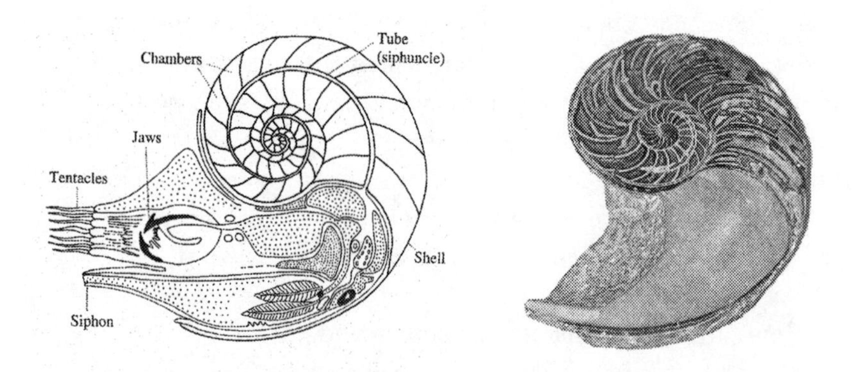

Figure 2.2 Diagrammatic cross-section of a living nautilus (eye not shown) and photograph of a fossil ammonite (part of tube preserved near centre of shell).

Much is known about the characteristic jaws of the living relatives of ammonites such as the squid. I was first drawn to this type of mouth while trying to identify the culprits of a particular form of vandalism. Many electrical cables have been laid on the sea floor, sometimes at great depths. Recently a fault was reported in one such cable, a few centimetres thick, which lay off the Australian coast. It was eventually discovered that the cable had been completely severed. Marine scientists were presented with a section of the cable close to the fracture, and the cause of the fracture immediately became evident – vandalism of some kind and vandalism by something with a beak. The scrape marks in the half-centimetre thick, black plastic casing did not match any bite marks of fish, jawed worms, dolphins or any other animal – except the squid. Squids and their relatives possess hard beaks, similar to those of parrots. Eventually a museum specimen of a squid beak was found which fitted exactly the scrape-like bite marks. We had found our culprit.

It is well known that certain designs of beak are adapted to suit specific food items, in the style of birds of prey or the famous 'beak of the finch' that Darwin found enlightening. The shape of the ammonoid jaw, which is occasionally found as a fossil, indicates they fed on small prey, probably planktonic. So there would have been a need to make regular vertical migrations to follow the plankton – plankton regularly make vertical migrations today. But most ammonoids became fossilised at the water's edge, which led to the construction of virtual ammonoid environments with characteristically modest depths. Was this a true depiction, or was the evidence just too circumstantial? Sometimes circumstantial evidence can be compelling, but in this case more clues were required to substantiate the ammonoid environment, and evidence was found in the internal tubes of their shells.

It was discovered that the nature and properties of the internal tubes provide a simpler indicator of strength of an ammonoid shell than do the outer shell walls and chamber walls, which are often complex in shape. In turn, the strength data gave rise to depth data, based on the critical pressure principle. Eric Denton's work on a living nautilus provided justification for this projection. And the conclusion drawn for ammonoids? Many species inhabited waters down to at least 600 metres in depth. But this only intensified the problem as to why most ammonoids became fossilised near to the shore. And it was the death of an ammonoid that held the final solution.

A nautilus shell will, if its living tissues are removed, fill with water, become negatively buoyant and sink to the sea floor. This is how the deceased nautilus had traditionally been considered. But such a fate has proved to be unrealistic. This postmortem sinking was found not to be true for an animal that died with its soft tissues in place. In such a case, gases derive from the process of decomposition of the carcass, which soon expel water from the body chamber and inflate the decaying soft parts. Then, within a few hours, the dead animal will float to the surface. At this point the water and gas levels in the chambers other than the body chamber have remained unaltered since death. But after a couple of days, the decaying body and shell part company and go their separate ways. For the shell, the remaining water in the chambers leaks out via the internal tube. Then it is free to float on the ocean surface, like a coconut, until it encounters land. There it comes to rest, and there it may become a fossil. Here is the solution to the shoreline-fossil problem, and also the reason for such an extensive fossil record of the once common ammonoids. Indeed, if they did begin to sink after death, with natural levels of gas in their chambers, they would reach only as far as their critical depths before imploding. In which case there would be no discernible fossil for such an abundant species. The nautilus story was concluded some thirty years ago, but recently the case was re-opened. A new biological study has revealed a twist, and what emerged to be a crucial adaptation for the ammonoids.

I have already referred in this chapter to the idea of mass extinction. Every so often, the history of life on Earth is punctuated by mass extinction events. There have been several cases of mass extinctions, the most famous happening sixty-five million years ago, which saw the demise of the dinosaurs. But the greatest mass extinction event of all, present predicament excluded, was the momentous Permian extinction.

The Permian, like the Cambrian, is a period in geological time with boundaries defined by events recorded in the fossil record. At the end of the Permian, 250 million years ago, around 90 per cent of the species on Earth disappeared. And again, the rocks can be employed to provide an answer to the cause of this event, and Doug Erwin of the Smithsonian Institution has pieced the evidence together.

The pore counts in leaves inform us that carbon dioxide levels and global temperatures were high 250 million years ago, following a cooler spell. A sudden drop in sea level at the end of the Permian destroyed near-shore habitats and destabilised the climate. With the death of the abundant flora and fauna that once inhabited the coast came decomposition on a grand scale. Decomposition results in carbon dioxide production, and, as the leaves predict, the carbon dioxide entered the atmosphere in significant amounts. This contributed to global warming and a depletion of oxygen that could dissolve in water. Unfortunately for Permian life, another disaster struck simultaneously – immense volcanoes erupted relentlessly for a few million years. To begin with, the eruptions cooled the Earth, but in the long term they led to global warming and ozone depletion. The effect of all of this on the oceans was that the water had become extremely anoxic – dissolved oxygen was scarce. It is therefore not surprising that most marine species became extinct; they probably suffocated. The filter-feeders were particularly hard hit, and the last of the trilobites disappeared for ever. Although many species of ammonoids also vanished, the ammonoids in general were among the few lucky ones – they made it through the Permian–Triassic boundary. How did they do it? This is where the new biological work on nautilus enters the story.

Recent studies have revealed a further adaptation in a nautilus living in deep water – its shell can behave like a Scuba tank. In the deep, oxygen levels can be low. It is well known that nautilus can counter this by lowering its chemical activity – it simply slows down. But it appears it also employs the oxygen in its buoyancy chambers to eke out the external oxygen supplies even further – it uses it to breathe. And the palaeontological story of ammonoids requires some adjustment because the Scuba scenario has been applied to the ammonoid shell. It is emerging that their Scuba tanks probably carried the ammonoids past the great Permian frontier. The ammonoids were highly adaptable when it came to levels of dissolved gases, and this probably accounted for their dominance throughout a prolonged period of history. The fact that nautilus continued with the Scuba system until today is good evidence that it indeed provides a competitive edge. So all in all the ammonoids were the master plankton fishermen – they could follow plankton everywhere, within a depth range that no fish today could hope to emulate.

This story illustrates the importance of understanding ancient environmental conditions before reconstructing ancient animals themselves. Suddenly all the fossil evidence of past climates and gaseous conditions is becoming relevant. But it is worthwhile also considering a different type of fossil evidence, one that can have equally important implications for fossil reconstructions. Ammonoids spent their life suspended in the water column. While alive they never set foot on the ground, their ground being the sea floor. Fortunately for palaeontologists, many animals did move on ground, and they left signs of movement in their wakes.

Trace fossils

Sherlock Holmes, and indeed his creator, Sir Arthur Conan Doyle, had a keen eye for footprints. Holmes used the size and type of print as an identification tool, the orientation of the prints to deduce entry or exit, and the spacing of the prints to determine the impetuosity of the crime. Palaeontologists, it seems, have converged on this practice.

Dinosaurs left their footprints in mud that became hard-baked and preserved through time. Today the prints are known as trace fossils – not parts of the ancient animals themselves, but impressions made by their movements. Footprints have revealed many secrets of ancient movement, feeding and lifestyles, such as group behaviour. This is all old hat. Now the study of dinosaur footprints has advanced a stage further, following the recent discovery of 200-million-year-old, three-dimensionally preserved tracks in Greenland.

In 1998 an American scientific team set out to explore the tree-barren fields of east Greenland. The team, which included Stephen Gatesy, Kevin Middleton, Farish Jenkins Jr and Neil Shubin, had been lured by the Triassic (over 200-million-year-old) exposures and the prospect of discovering early mammals. But the bones and teeth of various ancient vertebrates were temporarily cast aside as the team's attention became drawn to strange trackways of indistinct footprints.

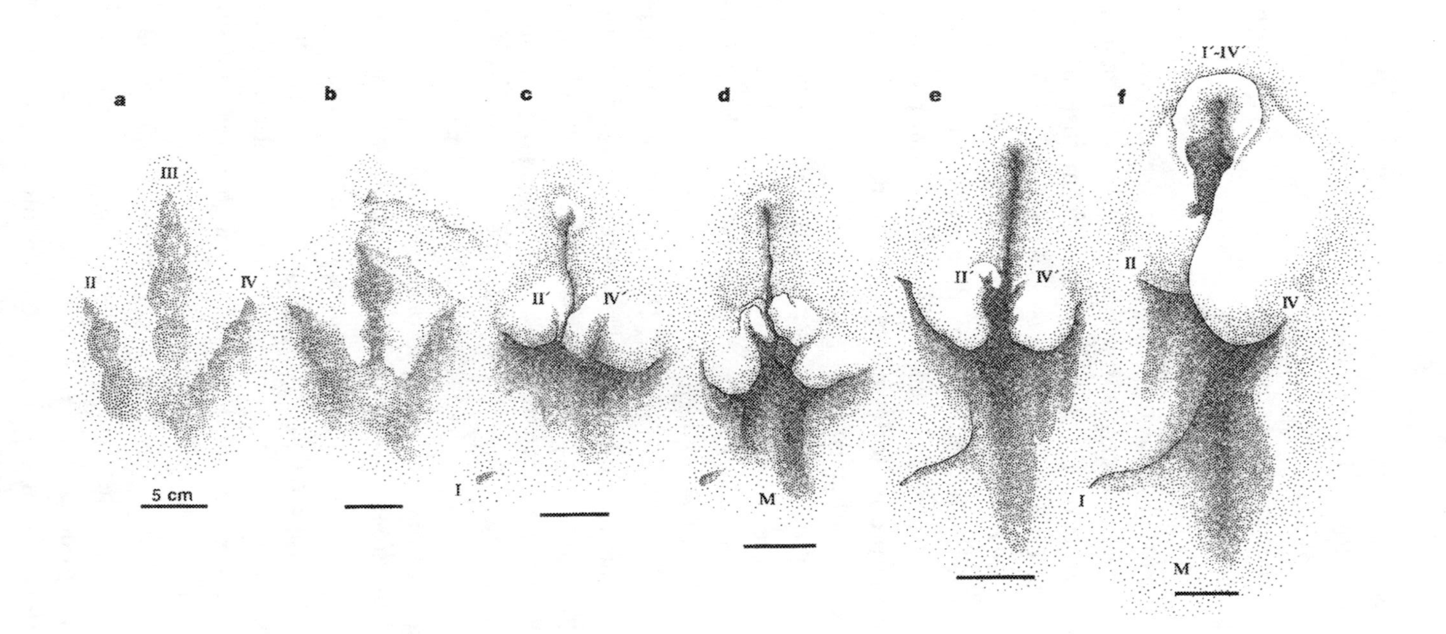

Figure 2.3 The footprints discovered in Greenland, made in both firm ground and sloppy mud.

There is a law among footprint workers: a trackway is not simply a record of anatomy. Rather it is a record of how a foot behaves under a particular pattern of movement as it makes contact with a particular type of ground. The varying conditions of ground can have a substantial effect on the features of the footprint – contrast a human print left in firm soil with one in wet mud. The Greenland tracks ranged from clear imprints to virtually indistinct traces, but they were made by the same species of theropod (carnivorous) dinosaur in, importantly, different types of ground. The ground varied from firm to sloppy, the range we find on a beach when we walk towards a fluctuating waterline. The prints made in the firm ground were run-of-the-mill, two-dimensional types as known from all corners of the prehistoric globe. As usual, they provided useful information about the owner of the prints and the precise form of the foot. It was the prints in the sloppy mud, however, that led to a breakthrough.

The sloppiness of the mud had preserved a three-dimensional footprint. It preserved the entry and exit 'wounds' made by the foot. And following comparisons with living animals, it transpired that the deeper you sink, the more of the movement that usually takes place above ground can take place below it instead. This was an important finding. It indicated that the three-dimensionally preserved footprints, regardless of their futile patterns at the surface, could potentially provide data on the movement of dinosaur feet through the air. Quite amazing when you think about it. And the only way of extracting the informative data was to examine the footprints in cross section.

The American team cross-sectioned plaster casts of the fossil prints in abundance. Eventually they assembled complete three-dimensional images of the footprints on their computers. But at this stage the three-dimensional prints appeared just as puzzling as the surface patterns. To make sense of them, the team turned to biology and studied living guineafowl and turkey. Live birds were run through increasingly sloppy mud and it became apparent that they left very similar, three-dimensional footprints. But it was the way they made the prints that was interesting and led to a theory of how dinosaurs moved their feet in the air as well as on ground.

Live guineafowl and turkey placed their feet into the mud with toes apart. But as they pulled their feet out of the mud their toes were brought together. When the birds walked on hard ground rather than soft, the same series of events took place, although this time they happened in the air. The same was concluded for dinosaurs – they opened their toes as their feet were placed on the ground, and closed them as their feet were lifted. Previously it was believed that some dinosaurs walked on the soles of their feet. But the sloppiness of the sediment revealed that in this theropod dinosaur the heel was carried the lowest, just a bit lower than in birds today. This in turn provided evidence that, compared to birds, the theropod stride was more strongly powered by the femur, while the lower leg and foot provided more of the power thrust.

The entire three-dimensional movement of a theropod foot through mud was modelled on the computer. This involved grafting the anatomy of a typical theropod foot on to the footfall pattern of a live bird. The images, and consequently the surface patterns made by the theropod, were self-explanatory (see Plate 7). It was nice to demonstrate that theropods walked in a similar fashion to birds, because the evidence from two-dimensional footprints and the bones themselves had been hinting at quite major differences in foot skeletons between the two groups. And of course this continued to feed the debate as to whether or not dinosaurs *were* 'birds'. Now it could be demonstrated that locomotion and limb function could have evolved gradually from theropods to birds, in common with many other features.

Adding further flesh to the bones

The precise relationship between dinosaurs and birds is a highly controversial issue. Signs of early feathers on a newly discovered Chinese dinosaur have been rejected by many, who prefer the interpretation that the downy outlines of the fossils are simply fibres from the skin that can fray when reptile skin surface is damaged. Ironically the specimen in question, a 120-million-year-old *Sinosauropteryx*, a theropod, has been brought to virtual life only to deliver a blow to its excavators, who sit within the 'dinosaurs-are-birds' camp.

The fine silt from an ancient lake had preserved the soft structures of *Sinosauropteryx*, including a clear silhouette of the lungs. John Ruben, a respiratory expert from Oregon State University, took one look at the 'lungs' and knew what he was dealing with. He had seen this lung arrangement before – in crocodiles. Immediately he constructed his virtual, living dinosaur, with the same compartmentalisation of lungs, liver and intestines that one would find in a crocodile, and not in a bird. This virtual dinosaur was incapable of the high rates of gas exchange needed for warm-bloodedness. So it contained cold blood, like the crocodile. Also, its bellows-like lungs could not have conceivably evolved into the high-performance lungs of modern birds. But still this evidence, that birds were not descendants of dinosaurs, is far from conclusive. As new fossils are unearthed and analysed with the lives of modern animals in mind, the building of a virtual dinosaur continues.

A study of vocal cavities and the surrounding bones has revealed the range of sounds once made by dinosaurs, from the high-timbred, lion-like roar of *T. rex* to the bellowing *Diplodocus*, with a voice reminiscent of air being forced from a hydraulic piston the size of a drainpipe. The nostrils of *T. rex* have been shifted further forward in its head to take a new position just above its mouth. Now *T. rex* has a much larger area of nasal tissue, fully laden with the capacity for a considerable sense of smell. This puts virtual prey in increasing danger, although as palaeontology becomes increasingly refined, maybe they too will become adapted, in this case to control their scents.

We identify the food of dinosaurs via the dentition of their jaws, the often fateful teeth marks left behind in bones, and their dung. But dinosaur dung has provided further information on ancient lifestyles and evolution – that of dung beetles. Radiating clusters of burrows have been found in Cretaceous dung that precisely match those made by dung beetles in elephant excrement today. These burrows indicate that dung beetles evolved with herbivorous dinosaurs, rather than with later occurring grassland mammals as previously thought. And here we have returned to the subject of trace fossils, which have breathed so much life into our models of extinct forms, right back to the Precambrian.

So dinosaurs are now running, breathing, smelling, roaring and excreting on our computer screens. The famous *T. rex*, whose skeleton was once constructed upright with tail on the floor, in the style of Godzilla, now lives its virtual life in horizontal stature – perfectly balanced with legs acting as a fulcrum. Similarly, *Diplodocus* no longer scrapes its belly on the floor. And if the makers of those first dinosaur reconstructions had taken note of the trace fossils, or consulted Sherlock Holmes, they would have noticed bold footprints but not a trace of lagging tails or hauling bellies in sight. Importantly and necessarily, dinosaur studies have led palaeontology well into the computer age.

Palaeontology meets modern engineering

More recently, the idea of producing three-dimensional models has been applied to fossils themselves. Travelling back some 400 million years, to pre-dinosaur times, certain marine organisms living in the shallow waters of the Earth were also preserved exceptionally well. Algae from these waters can now be found in New York State, some which have been replaced with pyrite but others which have been chemically unaltered and still contain their original organic material, like the flies mummified in amber. But more mysterious life forms of the era have been found. The exceptional preservation of these invertebrates has given rise to an unusual property of their fossils – they are three-dimensional.

The British team that recently discovered and began work on these fossils comprised David and Derek Siveter and Derek Briggs. The discovery itself was perhaps lacking in the romance of some better known examples. I pictured this research team flying within the Grand Canyon in a 1920s biplane, but my dream was shattered when I asked David Siveter about the locality of the fossils. He pointed to a large mound of earth visible from his office window even on a grey, rainy day. However, the ingenuity and excitement of this project lay with its methods.

During one decisive meeting, the research team examined the diversity of their fossils and realised that classification would be problematic. A view of only one surface or plane of a fossil, a view that fossils typically present, provided inconclusive evidence in this case, even at high magnifications. The three-dimensional preservation resulted in a limited view of the fossil, whose exposed parts lay flush with the rock. Imagine a golf ball embedded in a sand bunker with just one dimple exposed. The team knew there was more to these fossils than first meets the eye. To extract the maximum information, an unusual preservation called for unusual methods. In fact they chose to pioneer a new method for fossils. That method was risky – in the process of examination the valuable fossils would be completely destroyed. The gamble, however, paid off.

Today engineers employ computer-aided-design, or CAD, to construct and view car designs in three dimensions. Compared to pen and paper, CAD provides the advantage of enabling an object to be viewed in three dimensions and from all angles, as the object can be rotated on the computer monitor around any axis. The palaeontological team on this case wondered about the possibilities of introducing CAD to their analyses, and they soon enrolled a postdoctoral worker, Mark Sutton, with computer programming talents. But a hurdle lay ahead – the tiny fossils, perhaps only a couple of millimetres wide, required separation from the rock. Basically this was not possible for such a preservation type. So how could they determine the structure of all sides of the fossil – the food for the CAD-style program? This was the risky part – the fossils were to be ground away, a hair's width by a hair's width.

After each serial grind, a photograph was taken of the newly exposed section of the fossil. The palaeontologists were interested only in the surfaces of the fossils, since the innards had not been preserved. Although each grind revealed a redundant cross section, the photographs were fed, in order, into the computer, and the computer did the rest. The results were staggeringly good. In this chapter I have attempted to describe how fossils can be brought to life by piecing together one small fragment of evidence after another. Bit by bit fossils can grow virtual outer skins, fill with virtual blood, and walk across the computer monitor in search of specific virtual food. But in this case, what had involved years of work for other fossils happened in an instant. A complete animal, more than 400 million years old, came to virtual life on the computer monitor with one press of a button. The worm-like, armoured forms of early molluscs and segmented worms, some the earliest known representatives of their kind along with ancient arthropods, appeared exactly as they would have when they originally roamed the reefs. There were no fragments of anything, just the entire animals. And the 3D images could be rotated on the computer monitor revealing views from above and below, from the front and the back . . . from any angle one desired. Amazing! It is to be hoped that this CAD-style methodology will enjoy a happy future in palaeontology.

Taking our tools to the Cambrian

Now that my palaeontological tour has entered the twenty-first century, it is safe to return to the Cambrian. Chapter 1 referred to the exceptional preservation of the Cambrian Burgess Shale fossils. The preservation of fine details has led to precise classifications. Limb parts and trace fossils are preserved in the Burgess Shale, and now these can be extrapolated to bring the fossils to virtual life. In fact the Burgess Shale and other Cambrian fossil assemblages have paved the way for wonderful ecological models to be constructed for Cambrian communities as a whole. Greatly exaggerated in size, the newly constructed Burgess scene within the Royal Tyrrell Museum in Alberta, Canada, includes a walk through a Cambrian reef where animals interact all around us, above and below. Here, the palaeontological techniques discussed in this chapter are brought to a crescendo as the Cambrian comes to virtual life.

It was the conditions of their burial that destined the Burgess organisms to make scientific headlines, and to star in the detailed Tyrrell model. A combination of an ideal clay substrate with the right cations, pH, and carbon content possibly engulfed the living Burgess organisms to preserve at least some of them in the wonderful condition in which we find them today. Original organic material from the Burgess organisms has been preserved in at least some cases. Nick Butterfield of Cambridge University demonstrated this fact with his delicate separation of organic parts from the Cambrian rock. Acid was used to dissolve the rock matrix, where the fossils remained unharmed and simply floated away in the solution. These separated parts will be examined later in this book.

The Tyrrell model affirms that, as details of the Cambrian fauna are revealed at finer levels, the business of reconstructing Cambrian scenes is getting increasingly serious. The pioneer watercolours depicting Cambrian reef scenes, that for decades graced the corridors of natural history museums around the world, are making way for the sophisticated work of palaeo-artists. The crowded aquarium-like scenes of the amateur 'windows into the past' are becoming museum pieces themselves (quite literally). New reconstructions portray detailed movement in three dimensions among natural, spacious environments, as scientific principles are adhered to. X-ray photographs are revealing muscle attachment sites on the skeletons of Cambrian animals. Just as muscles were added externally to the first-century Jewish skull in lifelike proportions, now they are being added internally to the limbs of Cambrian arthropods – animals with exoskeletons like shrimps. When a skeleton is given virtual muscles in correct proportions, it can move naturally on the computer screen. Trilobite antennae are becoming conceptualised as flexible structures that can fold under the torso, as the body plates glide over each other and the animal rolls into a ball when danger approaches. On extension of the body, the gill plates are now considered to hang down quickly from the vaulted exoskeleton and flap in a style optimal for breathing when it is safe to do so. If all members of a community are brought to virtual life in this way, interactions between individuals and even entire food webs begin to manifest themselves. And work on Cambrian fossils and their reconstructions is accelerating.

In the dozen or so years since Stephen Jay Gould wrote *Wonderful Life*, advances in Cambrian biology have been considerable. The once 'bizarre appearances of problematic species' have now been more closely linked to living species, following the discovery of new, intermediate forms that fill the gaps. The once mysterious long, thin bodies, of *Hallucigenia* and *Microdictyon*, with spindly legs, have now been placed within the velvet worm phylum of animals. Velvet worms exist today with thicker, worm-like bodies and stumpy legs. New velvet worms have been uncovered from the Burgess Shale which share some important characteristics, such as claws, with living species and with *Hallucigenia* and *Microdictyon*. So the evolutionary holes have been filled.

A variety of trace fossils have been found in Early Cambrian rocks. These include branching and spiralling burrows, and U-shaped and more complex migration paths through the sediment. The surface of the sea floor has preserved the trails and resting traces of creatures that walked and glided over the sediment. These are the footprints of animals with elaborate body forms and behaviours, including animals that were the first to *walk* on this planet, with tiny yet huge and historically significant steps.

The biological indicators of environment within the Burgess Shale imply a tropical reef setting. But today the Burgess Shale is found halfway up a mountain in a snow-covered part of Canada – the furthest one can get from a tropical reef. Now it is understood that mountains today were perhaps once marine reefs, as a result of movements in the Earth's plates. In fact we can construct a world atlas so accurate that it could have been used to navigate in the Cambrian, at the time the Burgess organisms lived.

So the Burgess animals inhabited a near-equatorial position on the globe, hence their tropical environment. Now we would appear to know almost everything about their private lives, although surely the next decade will prove to be just as enlightening as the last. Remember the lesson learnt from the president of the Inventors' Association and his wisdom? But today we have amassed enough information, from ambiguous and esoteric signs of life flattened in shale, to consider the Burgess fossils as living organisms interacting within an ecosystem. And interactions between individuals from different species will prove to be important later in this book.

This chapter has demonstrated how our pictures of life on Earth through geological time have been constructed. By moving back through time gradually, and filling the gaps along the way, we can be less fearful of reconstructing ecosystems from an epoch as distant as the Cambrian. The logarithmic-style time travel employed in this chapter perhaps settles the nerves all the more – to begin with, it's nice to make

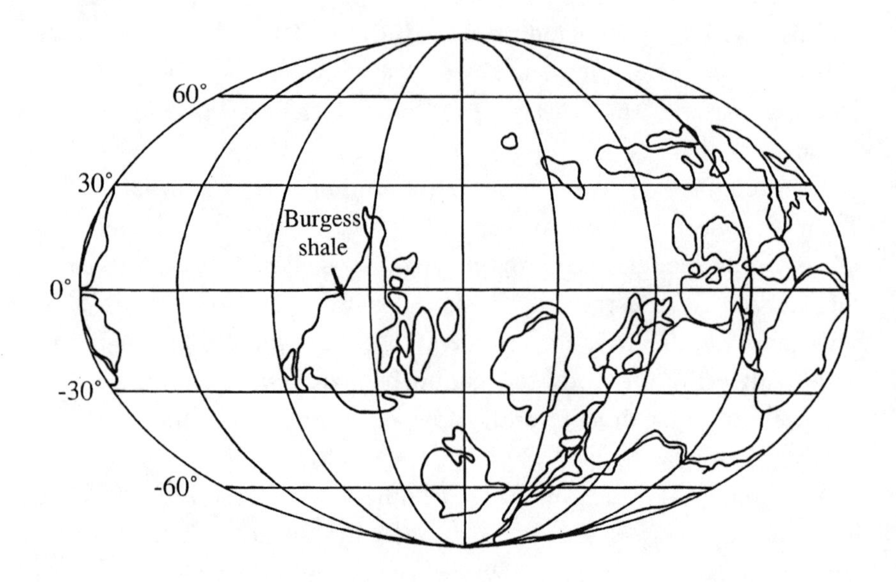

Figure 2.4 Palaeo-map of the world, at the time of the Burgess dynasty, showing the original location of the Burgess reef.

reconstructions of ancient but recent periods, periods we can test. Now we can be confident of the wealth of biological information we have extracted from Cambrian fossils, but this information ceases abruptly and simultaneously in all parts of the world as we split rocks formed before the Cambrian, beyond that auspicious borderline of 543 million years ago. It is no longer appropriate to extrapolate linearly and assume, with Darwin, that fossils of multicelled life with hard, external parts must exist in Precambrian rocks, implying that we simply haven't found them. Our fossil finds since Darwin's days have increased a hundredfold. But we are still without Precambrian signs of the characteristic external parts of animals today.

The characteristics of the fossil record through time have been assessed using quite convincing analytical methods, where the past 540 million years were examined. Although ancient rocks preserve less information on average than more recent rocks, the fossil record since the Cambrian explosion provides uniformly good documentation of the life of the past, and there is no reason not to extend this trend into the

Precambrian. So we are still without Precambrian fossils of the variety of body forms expressed in phyla today, other than the sponges, comb jellies and cnidarians. It seems certain that our modern view of animal evolution, and the Cambrian explosion, is correct. Equally, despite the numerous mass extinctions and recovery from those events *since* the Cambrian, new animal phyla have not evolved. These conclusions have been strengthened with every fossil discovery.

In addition to presenting the evidence for Chapter 1, it is to be hoped that this chapter has supported a statement from John Maynard Smith's *The Theory of Evolution*: 'The study of fossils . . . can be made to reveal the way of life of animals now extinct.' We have ascertained how animals ran, swam, flew and burrowed in previous times. We have deduced their feeding habits, their daily activities and their favourite pastimes (almost). But after all this detail extracted from the fossil record, after all the constructions of virtual lifestyles, virtual climates and entire virtual ecosystems, there is still something missing from our interpretations of the past – colour. Is this a serious omission? It is time to examine colour in life today.

3

The Infusion of Light

> Whenever colour has been modified for some special purpose, this has been, as far as we can judge, either for direct or indirect protection, or as an attraction between sexes
>
> CHARLES DARWIN, *On the Origin of Species* (first edition, 1859)

A series of Victorian doorways, staircases and corridors within Oxford University's Museum of Natural History eventually lead to the step of a more humble entrance in the far corner of the Gothic-style building. This is the door to the Huxley Room. Beyond this door lies an historic roof – its timbers absorbed the first words of evolution ever to be spoken to the public, during the Great Debate of 1860. Here, Thomas Huxley matched Bishop Wilberforce blow for blow in the original 'science versus religion' showdown. Huxley was defending Darwin's *On the Origin of Species*, published seven months earlier, in an attempt to prevent 'sentiment interfering with intellect'. Darwin himself was absent, but Huxley skilfully succeeded in his task, and evolution began its infusion into the global language. It is worth pausing at the door to the Huxley Room.

After the Great Debate, the Huxley Room became moulded into an entomological collection room – it was filled with preserved insects. The last Victorian curator of the insect collection at Oxford's Natural History Museum, Sir Edward Bagnall Poulton, became fatefully attracted to the beetles within.

Poulton opened the door to the Huxley Room one morning, and as usual he took time to appreciate the architecture. Streams of sunlight, illuminating the gently sloping sides of the roof and the many decorated beams, cut into the darkness of the room. He passed down the aisle of the Huxley Room, created by the two rows of wooden entomological cabinets. A drawer that had been removed from its cabinet for some time caused him to pause during his general inspection. The drawer was struck by the sun's rays, which were streaming through the round lens of a leaded window and became focused into a beam. Poulton blew away the dust from the glass lid and his eye, which had adapted to the intermittent darkness of the room, was at once arrested by a jewel. The beam of sunlight had ignited the metallic-blue colour of a carrion beetle, about the size of a thumbnail. The label, attached to the pin supporting the specimen, read '*Oiceoptoma*, Sumatra, Wallace 1866'. It was fitting that a specimen collected by Alfred Russel Wallace, a cofounder of the theory of evolution, should be found gleaming in a room where the theory had first been put on trial. Indeed, Darwin had collected other specimens here, but it was in the colour of that Wallace specimen that Poulton's real interest lay. Soon Poulton was placing all the entomological drawers under the beams of sunlight, which reflected rainbows on to the Huxley timbers, those pillars of evolutionary learning.

Poulton eventually published a classification of colour in animals, and became 'the centre of gravity of entomological research in the British Empire'. He inspired a century of research on animal colouration and, in some ways, the clues that can be sought from this chapter towards solving the Cambrian enigma.

Before the Victorians

Some millennia earlier, Egyptians spoke of the 'Sun God'. They elevated the dung beetle to the status of higher being as it symbolically rolled sun-shaped objects around the desert. This 'scarab' beetle was believed to represent the sun god Khepri, and in the Egyptian language the word 'kheper' means both scarab beetle and existence. The Romans shared a similar interest in sunlight, though not only in a religious context. Heliography is the Roman art of signalling using the sun's rays reflected on metal shields. It was sometimes employed to dazzle enemies, when sunlight was directed momentarily into their eyes. A flash of light is more conspicuous than a steady light, but at close range it can have a stunning effect. From an aeroplane, the reflection of sunlight from a car windscreen is extremely conspicuous, if not blinding. Unfortunately for the Romans, they were foiled by their own technology – Archimedes later engaged metal shields to concentrate sunlight on to the sails of invading Roman ships, causing them to burst into flames.

Exceptional intensities of light are also experienced in the natural world. If sunlight can be focused to cause materials to burn, imagine the effect it will have on a retina. And as the retina has been a product of evolution, so has retinal destruction. On this subject, angelfish become Hell's angelfish when territory is at stake.

Angelfish live in the clear surface waters of the Amazon. They have flattened bodies with silver skin, similar to a mirror. When one fish invades another's territory, the defender leaves the shelter of reeds to do battle. Battle stance is a tilted position in the water column, with the aim of firing sunlight into the eyes of the opponent. Like Roman shields, the strong Amazonian sunlight can be concentrated into a narrow beam and directed precisely. In fact both fish in this combat take up their positions in the open water, fine-tuning their lines of fire by adjusting the tilt of their bodies. Light flashes through the water like the lasers of *Star Wars* battles. The stakes are high. A direct hit in the eye can lead to the bursting of blood vessels and an increase in heart and breathing rates. A fish defeated in this manner is at best temporarily stunned and at worst killed. Either way, the battle is over. This is a fish living in waters where sunlight is at its most intense, and it has adapted. Acting on this strong selection pressure, it has evolved precision mirrors.

But what exactly is light? I am about to launch into the history of this key question. The question can in fact be divided into smaller inquests, the answers to which are staggered throughout the history books. Such a fundamental element of optics, however, may at first seem irrelevant to a book on evolution, even one addressing colour in nature. So why not simply provide the answer and spare the history lesson? There are clues to the cause and purpose of colour in nature throughout the historical accounts of scientific enterprise, and even throughout accounts of artistic and military endeavour. Human ingenuity and artistic expression have often converged on natural selection where colour is concerned.

In this chapter we will observe a specific animal and ask two questions: 'What causes *that* colour?' and 'What is the purpose of *that* colour?' We will move through a list of animal species, providing different answers to these questions. It will emerge that there are many possibilities in each case, but the triumphs and even tribulations of early scientists, artists and military tacticians can help narrow down the list of suspects.

As others did before him, Leonardo da Vinci strove for an explanation of light in the fifteenth century. He did, however, see things a little differently from his predecessors. Leonardo began to doubt the philosophers of his time, who believed that light was something emitted from the eye, returning on reflection from an object – the object observed. Light, Leonardo believed, was comparable to sound, and both travelled through air or water as a 'tremor'. By this he implied a signal which is spread via a sequence of disturbances in the air or water – he was describing a *wave*. Throwing two stones into a river, and observing the corresponding sets of concentric waves break each other up, Leonardo wondered if light behaved in a similar way.

Leonardo became distracted and turned his attention from light to everything in the cosmos. He suggested that 'everything was propagated by means of waves'. He should have stuck with light. At least Leonardo came to the conclusion that light was a property of the sun. From now on philosophers could think of light in terms of waves, albeit in their simplistic form.

The wave idea was taken a stage further by, and often credited to, both Christiaan Huygens and René Descartes. In 1664 Descartes described what happens to light as it passes through raindrops. He concluded that internal reflection caused the effect of a rainbow. But light was in theory only white at this stage in history. And so, consequently, was the predicted rainbow. Descartes also believed the propagation of light was instantaneous. He was to be proved wrong on both accounts.

Later in the seventeenth century, the French mathematician Pierre de Fermat breathed new life into Leonardo's ideas that nature always acts by the shortest paths, and that light *did* travel at a finite speed. And according to Fermat, light travels at different speeds in water and air.

At around this time, the twenty-two-year-old Isaac Newton was discharged from Cambridge, interrupting a Bachelor of Arts degree to escape the Great Plague, which was making its way towards his university from London. The two years that followed saw perhaps the most creative display of individual genius in the history of science. At his home in Woolsthorpe, Newton formulated the binomial theorem and the differential and integral calculus in mathematics; the unification of celestial mechanics; the theory of gravity in astronomy; and . . . the theory of colour in optics. In what came to be called his *experimentum crucis*, Newton split light into a spectrum of colours using a prism. Then he passed each 'colour' through a second prism to demonstrate that further fractionation was not possible. Newton had shown that sunlight was the combination of the complete colour spectrum, and no more. Now Descartes' rainbow could have its colours.

Newton did not have a strong view about the nature of light. In fact he favoured the idea that light consisted of particles, where the particles of different colours had either different speeds or masses. But Newton found no time to test this notion with his usual high degree of mathematical exactitude. It was Huygens' wave theory of light that ultimately triumphed (although today we consider that all particles can behave like waves, and vice versa).

In 1690, Newton's contemporary Christiaan Huygens stated categorically that each point in a wavefront is the source of new waves with the same frequency of oscillation. Wavelets can be cancelled out by other wavelets, travelling in a different direction, like the opposing ripples travelling from Leonardo's two stones. But in the absence of obstacles, the wave progresses forwards.

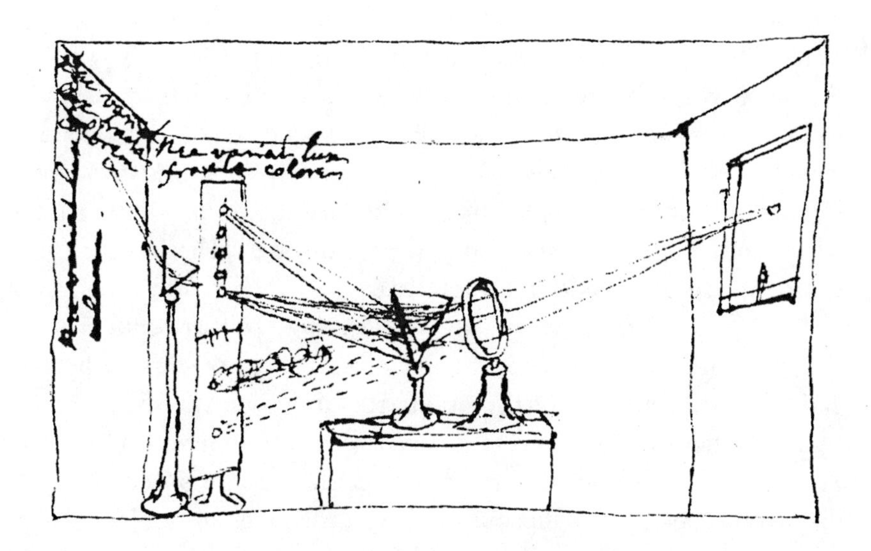

Figure 3.1 Newton's own drawing of his *experimentum crucis*. Unfortunately he lacked the artistic genius of Leonardo.

Another Victorian curiosity

The Victorians of the nineteenth century were handed all of this knowledge. They knew that sunlight contained waves of different wavelengths, and that each wavelength could be converted to a different colour by the eye (colour does not exist in the environment, only in the mind – this will be discussed in Chapter 6). But the Victorians had precision apparatus at their disposal, and on the characterisation of light they finished off what Leonardo had begun. That is, they finished it off for the purpose of this book (apologies to Planck and Einstein).

The early Victorian English physicist Thomas Young found that any colour could be obtained by combining only three different colours – blue, green and red, a useful concept that became important to science and television. Then Young discovered polarisation. When a wave travels along a guitar string, the displacement of the string is sideways. If a narrow slit is introduced into the path of this 'wave', the wave will continue only if the displacement is parallel to the slit. If it is not parallel, the wave will be reflected and will travel back on itself. Light behaves similarly – it is a transverse wave. Polaroid sunglasses approximate a slit to reduce light transmission. If a beam of light contains waves with different directions of displacement, or polarisations, only those parallel to 'slits' in the lens material will pass through. The light passing through each lens is said to be polarised.

Meanwhile, Victorian scientists were tackling another conundrum – the speed of light. Previously, the tiny displacement of stars in the sky caused by the Earth's orbital motion around the sun was exploited to reach a surprisingly accurate value. But in the nineteenth century French and Polish scientists aimed to measure the speed of light directly, which required ingenuity and Victorian high technology. They conducted experiments where a rotating mirror was illuminated by a lamp, creating pulses of light. A second, static, mirror was placed at great distance in one direction. This reflected one light pulse back towards the rotating mirror. As the rotation speed was varied, the returning light struck the rotating mirror at slightly different angles. At one angle only, the light was reflected back towards the lamp. Using the rotation speed of the mirror, and the distances and angles involved, the speed of light was calculated, quite accurately, as 186,355 miles per second. So light takes about eight minutes to reach us from the sun. This fact became united with the work of Maxwell.

The Scottish physicist James Clerk Maxwell became most famous for his theory of the electromagnetic field. To cut a very long story short, Maxwell found that electric particles in a medium such as air are displaced from their normal positions by the action of an electric field.

Maxwell came to realise that electric particles in his experiments were displaced through the medium in wave form. But he was also able to calculate the speed at which these waves travelled – it was the same as that measured for the speed of light! Eureka! Maxwell had discovered that light is in fact electromagnetic waves. That is, they have an electrical component and a magnetic component – waves with perpendicular displacements. And in the 1880s the German physicist Heinrich Hertz confirmed Maxwell's theory with some ingenious experiments.

But how does all this relate to the most famous Victorian science book of all, *On the Origin of Species*? Evolution has been subjected to the principles of optics, too.

Pigments

Travelling home on the Manly ferry one night, within Sydney Harbour, I observed Young's theory of colour mixing in action. Part of the harbour is fringed with skyscrapers of varying heights, but all displaying their company names in neon lights. The lights are reflected from the water, forming a mirror image. But I noticed that some colours in the reflections were absent from the neon signs. Ripples in the water surface were mixing the reflected images of different buildings, including their neon signs. Where red and blue signs lay above each other on the horizon, I saw only a purple reflection.

This principle of 'effectual colour mixing' – as opposed to simply mixing different coloured paints on the palette – was a favourite of the French Impressionists towards the end of the Victorian era. Camille Pissarro's painting *Peasants' Houses* shows clearly a peasant passing through a garden gate, in front of a row of country cottages. Move closer to the painting and the scene becomes pixillated. Suddenly the houses disappear to become a collage of red, blue, green and yellow streaks. At distance the red and blue merge to reveal the purple shadows of chimneys – the red and blue streaks can no longer be separated by the eye. Adjacent red and blue streaks fall into the same pixel of our eye's picture. And convergence exists in nature.

The Atlas moth grows to the size of a standard dinner plate. Its huge wings incorporate patterns of mustard and grey. This colour derives from pigments in the scales. Pigments, like those in artists' paints and our clothes, are molecules that absorb certain wavelengths in white light. These wavelengths are no longer available to vision, but the remaining wavelengths in the sun's spectrum are reflected from, or transmitted through, the pigment system. These are the wavelengths we see. And this is the commonest cause of colour in animals and plants – they contain pigments.

Actually, the mustard and grey Atlas moth has no mustard or grey pigments in its wing scales. Place the wing under a microscope and the grey areas become a mixture of black and white scales, and the mustard areas a mixture of brown and yellow scales. Examining the moth's pigments at another level, in mustard we have two different types of colour – saturated and unsaturated. The yellow is saturated – that is, it contains the wavelengths for yellow and that's all. This is saturated in the Newtonian sense. If a slit is introduced into the path of the spectrum separated by Newton's prism, so that only the yellow part can pass through, we have a saturated yellow. Brown, on the other hand, is an unsaturated red. If the slit is moved through the spectrum so that only the red passes through, and that red is diluted by the addition of a faint white light, then brown is the colour observed. So brown is said to be unsaturated because it contains a broad range of wavelengths.

Most of the colours on the Great Barrier Reef are the result of pigments. It's great to see such amazing colour *and* understand why it is so. One of the aims of this chapter is to explain how this understanding comes about; with a basic understanding of the cause of colour, one can pass through any environment and explain the hues of all its inhabitants. Although there are many other ways to appear coloured, as will be revealed later in this and other chapters, each mechanism has its own unique signature in terms of optical effects. Pigments can cause an animal, or part of an animal, to be coloured, but this colour is not the most dazzling type. Also the colours caused by pigments do not change with the direction of viewing, or when the animal itself moves. This is because pigments disperse or reflect wavelengths equally in all directions. They will thus look the same from every direction, over a complete hemisphere. And because we see only a very small cone of that hemisphere at any one time, because of the small size of our eyes, then we can receive only a small portion of the wavelengths present in the sunlight. If we possessed eyes the size of footballs, pigments, especially when close up, would appear much brighter. So the light we see is much dimmer than that of the original illumination – sunlight. And as we move further from the animal in view, the cone of light detected becomes smaller and the light dimmer. Eventually it will fade to below the limits of detection. Consider a landscape fading in the distance.

Back on the Barrier Reef, as described at the beginning of this book, the cuttlefish's brown ink contains a pigment released into the water. While following the cuttlefish, the corals I passed over had as great a diversity of colour as forms. The saturated reds, yellows and oranges looked the same from all directions, indicating that pigments were behind them all. The sponges covered the complete spectrum, and like the reds of anemones, lobsters and starfish, they bore the saturated coloured effects of pigments. The purples and browns of sea urchins indicated unsaturated colours. But as I followed the cuttlefish around them, again their colours did not change, indicating pigments all the same. Then, as I mentioned previously, something happened to the colour of my guide. The otherwise brown and white cuttlefish turned red . . . then green.

Take a close look at a colour TV screen. When it is switched on, clusters of blue, green and red 'sub-dots' are distinguishable. Each of these sub-dots continually, and independently, becomes brighter and dimmer as the overall picture on the screen changes successively. This, again, is Young's colour mixing in action. Black and white photographs in newspapers are constructed from regularly spaced black dots on a white background, just like the shades of grey achieved from the black and white scales of Atlas moth wings. The size of each dot determines the shade of grey in its particular region. The picture on a TV screen is constructed from dots, too. But here a comparable dot is made up of three sub-dots – one green, one blue and one red. And by changing the brightness (rather than size) of each sub-dot, the overall dot can appear any colour of the spectrum. So as a yellow tennis ball flashes across the grass court on the TV screen, different combinations of sub-dots glow. When the ball is over a dot, the green and red sub-dots light up, while the blue is off, to produce yellow. As the ball passes, the red sub-dot also turns off to leave green. And a yellow wave of colour travelling across a green cuttlefish works in the same way. But how can this be? Pigments produce permanent colours; they cannot suddenly change. The leopard, for instance, *cannot* change its pigmented spots. It was the Victorians, again, who made sense of this paradox, although not at the first attempt.

In 1802, Tom Wedgwood, son of the potter Josiah Wedgwood and uncle of Charles Darwin, took one of the first forms of a photograph. He painted leather or paper with a solution of silver nitrate, which is sensitive to light. He placed leaves on top and exposed the apparatus to sunlight for about half an hour. The light turned the exposed silver nitrate to silver metal, which reappeared, and the shape of the leaves emerged as pale silhouettes. A negative had been made, albeit in black and white. So of course colour photography became the next great Victorian goal, one eventually achieved by James Clerk Maxwell.

Prior to Maxwell's accomplishment, the nineteenth-century scientist Otto Wiener believed the colour breakthrough lay with compounds of silver chloride that react with different wavelengths of light. The new compounds formed at the end of the reaction would have the colour corresponding to the catalysing wavelength. Wiener also thought that organic substances, such as those found in animals, could possess a similar property. Then came a theory of adaptive camouflage. A caterpillar, Wiener argued, might vary its colour to match a changing environment because its skin 'photographs that environment by means of the sensitive compounds of its own tissues'. A nice idea, but pure fiction.

The eminent Victorian naturalist Henri Milne Edwards made amends in 1848. Like Aristotle, and philosophers, scientists and poets since, Milne Edwards was intrigued by the chameleon. Chameleons change their colour dramatically. The big question is, 'How?' Milne Edwards realised the answer lay not with any chemical change in the skin, but with the mechanical distribution of pigments. This was a breakthrough.

The skin of the chameleon or cuttlefish is packed with chromatophores – colour cells. These are simply cells packed (usually) with pigment. Each colour cell contains just one type of pigment that causes one colour. But the cell is elastic – it can change its shape. Under nervous control, it can become flat and thin, lying parallel with the surface of the animal, or short and squat. And the pigment is spread evenly throughout the cell in each case. Looking at the animal, the short, squat cells reveal only a small area of pigment, and the visual effect is negligible. But the thin, flat cells reveal much more of their pigment, and can be seen by the naked eye. Compare these two possible forms of the colour cell, considered off and on, with a coin. A coin is easily observed when lying flat, but it is more difficult to see edge on.

Chameleon and cuttlefish skin is actually packed with colour cells of various hues. In comparison with a TV screen, individual cells can be considered sub-dots, collectively forming dots that can independently cause any colour. By being turned on and off, or by becoming an intermediate phase, the different sub-dots contribute to a dot that is capable of assuming any colour of varying brightness. At high magnification, imagine the skin as an assortment of juxtaposed and coloured coins. When some coins are turned on their sides, different overall colours are achieved. And this works – it really is extremely effective. One would hope so, too, considering the evolutionary trouble involved and the physical costs of such a mechanism. Significant electrical wiring, brain space, production of pigment and specialised cells, muscles, and sensors are required. With these costs in mind we can begin to consider the importance of light as an evolutionary factor and behavioural concern. The importance of this cannot be overstated.

Evolutionary interlude

If an animal does not adapt to the light in its environment, it will not survive. Today light could be considered the most powerful stimulus in most environments on Earth. In this chapter I will continue to demonstrate this point, using examples of how the world we see is one adapted to light. I do not intend to diminish the significance of other stimuli, such as touch, sound and chemicals, for these are hugely important, too. But light is an exception among stimuli because it is always there. If you don't make a scent, you will not be smelt. If you don't make a sound, you won't be heard, although for some animals silence and lack of scent are difficult to achieve. Touch is a little different because it operates, obviously, only over very short distances. The adaptation to light is a vital necessity. Light is where the sun's radiation peaks. It exists in many environments on Earth. If it did not, life today would be very different.

There are a couple of exceptions to this rule of exclusivity. Two other stimuli exist in the environment that also cannot be avoided. Many bats hunt using radar. They produce pulses of ultrasound that return to the bat after rebounding from an object, just like the military radar system that detects aircraft. If, at night, the bat's radar detects an object that is small and in mid-air, it is probably a moth. That's food to a bat. But just as animals living under the sun are adapted to light, so moths are adapted to radar. They are covered in a sort of radar-absorbing fur, which reduces the signal reflected back towards the bat. When the radar source is very close, they can stall and dodge the oncoming bat. A similar cat-and-mouse game takes place underwater, where dolphins hunt fish using a comparable stimulus – they produce sonar.

Also in the water, some fish produce a different stimulus. Electric fish such as the numb ray and electric eel were once targets for those who doubted evolution. How could such a strong, complex and specialised characteristic suddenly appear in the history of animals, as if out of nowhere? Any evolutionary shudders were stilled on the discovery of the 'missing link' – weakly electric fishes. These fishes do not produce the high voltages capable of killing prey by their mere touch. Instead, weakly electric fishes emit faint electric fields that work in a similar way to sonar. They can select prey based on the electrical signal that is returned. And from this the strongly electric fish could evolve.

Radar, sonar and electric fields, however, are rare on the surface of the Earth in comparison to sunlight. To begin with, an animal must produce its own stimulus, although this is sometimes worthwhile because, like light, it becomes a stimulus that other animals cannot avoid without taking action. Stimulus production is an expensive exercise all the same. So the fact that it exists in nature indicates that it does work, and works well. But still the environments that carry these stimuli are very limited. Also, sonar and electric fields only affect animals of a very specific size – the size of food for the stimuli producers. Yet with light, there is always an animal, or more realistically many animals, which will have an interest in the optical signature of *every* animal living under sunlight.

So animals have to accept, or in evolutionary terms *adapt* to, the sunlight that strikes them. There are two routes an animal can take – the path to camouflage or the path to conspicuousness. At the foot of this evolutionary junction, the balance may be even. The path to take could be purely under the influence of chaos. It could also be influenced by the materials available for evolution – the building blocks, or atoms in the case of pigments. But, as will be demonstrated in Chapter 5, once the balance has tipped one way, evolution can continue full speed ahead along its chosen path, until there's no turning back. And it is this balance of camouflage ('indirect protection') and conspicuousness ('direct protection or attraction between sexes') to which Darwin referred in the epigraph at the beginning of this chapter.

The purpose of pigments

When the Australian colonists entered the mountainous terrain of Papua New Guinea in the 1930s, they were amazed to find some of the population still in the Stone Age. Tribes there lived under a cyclical regime of peace and warfare.

Until the late 1980s, battle in New Guinea involved spears, arrows and shields. Shields were carved from tree trunks and were often as tall as their owners. These shields were painted with locally available pigments, in geometric designs. Anthropologists made early attempts to interpret these designs, but they were on the wrong track. The designs carried no meaning; they were there simply to intimidate the enemy. Indeed, the warriors also painted themselves, making them 'glint terrifyingly'. The overall bearing and brilliance of a warrior with his shield warned of his support by ancestral ghosts . . . and this was backed up by a large spear. The pigments were warning colours advertising the threat posed by the warrior. In this context, his weapons were also ornaments. Warrior colours may have incited surrender or retreat before battle had chance to commence.

Following the decommissioning of armour, European armies employed warning colours up until the nineteenth century. The bright red and white uniforms, with tall headwear, provided a warning message or two. Like much of the armour before, a large hat provides a false impression of body size. The larger the individual, the greater the threat perceived. And the immaculate dress itself was a clear symbol of a well-disciplined army. Then, of course, there were the regimented manoeuvres. This was an army that was prepared and knew what it was doing, at least in the eyes of its enemy.

During the nineteenth century the philosophy of battle colours changed. With the introduction of accurate, long-range guns came a new form of advantage for the soldier.

Until this time, although conspicuousness had been the soldier's battle principle, there was always an alternative lurking in the back of the brigadier's mind – camouflage. Merging into his surroundings, a soldier could either avoid or surprise the enemy. But then armaments really would be armaments, and the enemy would be fearless. Ornaments would become obsolete. So there was always a balance within military intelligence, just like the balance within nature, between conspicuousness and camouflage. And the military balance eventually tipped the other way.

New weaponry called for new tactics. Armies fought at greater distances apart – so far in fact that the smart uniforms, never mind their shiny buttons, were simply not visible. Although the regimented formations continued to instil some degree of fear, in general it was fading, like the pigments themselves over distance. Now the bright red uniforms served only as targets, and the path to camouflage became the route to take.

The balance between camouflage and conspicuousness lies behind every case of purposeful colouration in nature. Whether the colour seen is conspicuous or inconspicuous indicates the way the balance has tilted. This is the direction of evolution – the direction with the greatest difference between positive and negative selective pressures.

Dropping the military metaphor, the employment of pigments to provide an 'attraction between sexes' is a simple and straightforward concept in nature. Many obvious examples could be listed. Think of the birds of paradise, with their dull females and flamboyantly costumed males. Then there are the male hornbills that actively wear alluring (to a female hornbill) yellow make-up, secreted from preen glands and applied to their wings by the bill. But the other functions of colour as listed by Darwin are equally bountiful in nature.

Pigments are employed to provide 'direct protection' through advertising. The unicorn fish inhabits Hawaiian waters. Its name derives from a single, horn-like protrusion from its head. But another obvious characteristic of this fish is a strong spine on either side of its tail. The spines have a protective function – they can potentially slice open an aggressive fish with a single swish of the tail. And they are made obvious by their bright yellow pigments – a warning not to disturb this species. The warning is heeded well and the fish is left alone. The armaments are, again, ornaments.

Pigments may provide 'indirect protection' through camouflage. The peppered moth provides the case that first springs to mind. This well-known species is, as seen in its seventeenth-century guise, a pale grey colour so that it can camouflage itself against the silvery bark of trees as protection from predatory birds. During the Industrial Revolution, trees growing near factories became blackened by smoke pouring from factory chimneys. The pale grey moths were suddenly conspicuous against the black trees . . . or they would have been if it hadn't been for evolution. As selective pressures changed, new genetic mutations became advantageous – the ones that coded for black pigments. Thus the peppered moth became black in industrialised areas – its camouflage was restored. The moth had adapted to its new light environment, and it survived there.

Unfortunately for some other moths, their camouflage code is all too often cracked. But the moths are prepared for this. In the event that their cover is blown, they opt for conspicuousness as a last resort. The camouflage of these moths is confined to their upper wings – the only wings visible during rest. But when danger comes too close for comfort, their lower wings are quickly displayed, along with their warning colouration. Predators are confused by these unexpected blazes of bright colour and, in theory, the moths buy some time to escape. 'Flash' colouration is employed commonly by camouflaged animals, and so it must work . . . so long as the predators' approach is detected.

A variation on regulation camouflage is disruptive colouration. The tiger's stripes and giraffe's patchwork patterning break up the outline of the animals themselves against their natural backgrounds. Then at times they may provide good old regulation camouflage. Sometimes repetitive patterns are less noticeable than a continuous, albeit camouflage, colour against a busy, varied background. Closely packed trees provide vertical lines with leaves of different colours, shapes and, according to Pissarro, finely pixillated patterns. This situation calls for equally busy camouflage patterning, and the precise colours may be less important.

Outside Sydney University, there is a large pond full of water lilies, complete with lily pads. Admiring the plant life there one day, it was some time before I realised I was also watching a large black and white bird. But how could this be? The bird was black and white against a background of green leaves – surely the bird would be conspicuous?

Although green, the lily pads were also curled and shiny, and where they reflected sunlight into my eyes they appeared white. Standing on the lily pads the white patches of the pied bird matched those of the reflections from the leaves. So the white areas of the bird were removed from possible conspicuousness. The remaining black areas of the bird should, in theory, have been obvious against the green leaves. But they no longer formed the shape of a bird . . . or anything recognisable as such to me. And the bird itself had escaped my attention. I learnt that having more than one colour can provide camouflage even if only one of those colours matches the background. And another lesson learnt was to consider nature's colours only in their natural environments. The green leaves would, in the laboratory, have appeared a continuous green colour, against which the pied bird would have been quite prominent. This was not the case in the natural environment, under bright sunlight.

Monet provided a warning that one should beware the fixed, stereotypical image of an environment. He painted most of his landscapes many times, but at different times of the day . . . and his paintings were all unique. The epitome of this concept of immediacy is recorded in two of his haystack paintings of 1891. Painted at midday, the haystacks appear yellow, but in his evening interpretation the haystacks are glowing red. Under yellow light an object with a complete spectral repertoire will appear yellow; under red light it will appear red. To see this principle in action, try looking at the pages of this book under different light. The paper reflects all spectral colours, but under shaded sunlight it appears a bluish white, and under a light bulb a yellowish white. These are just two of the light conditions that call for different camouflage colours. So different constraints are placed on animals active at different times of the day, when different selective pressures are in action.

The Atlas moth has been considered so far under white light only. But under different colours, the moth assumes different appearances. Under red light, such as would be observed during the evening, the Atlas moth reveals patterns of stripes, providing disruptive colouration. Under green light, the moth exhibits a similar pattern to that under white light; that of regulation camouflage. So depending on whether the time is midday or evening, the Atlas moth sends out a slightly different message, albeit one intended to avoid the attention of predators in both cases. But there is more to this story. There is another colour contained within the sun's rays, just before violet in the spectrum. It is a colour that thwarted Leonardo, Newton and the Victorians, because humans cannot see it. That colour is ultraviolet.

Beetles and birds send secret messages written in ultraviolet through the atmosphere. We know this because their ultraviolet colouration can be recorded on camera film. Like the lenses in our eyes, glass absorbs ultraviolet wavelengths. Fix a quartz lens to a camera, however, and the ultraviolet transmits, and affects the camera film in the same way as violet or blue light. When this film is developed, we can observe the ultraviolet plumage of the budgerigar, for instance. But if we cannot normally see ultraviolet, why should we even consider it for biological purposes? Well, other animals, especially birds and insects, can see it.

Many flowers include ultraviolet in their colour palettes to attract pollinating insects. If birds can generally see in ultraviolet, and birds eat Atlas moths, it is important to know how the Atlas moth appears under ultraviolet light. Does it continue its camouflage or disruptive colouration into the ultraviolet? The answer is no. Under ultraviolet light the Atlas moth takes on a remarkable transformation. It appears as two snakes, with prominent bodies and heads, with eyes and mouths. The purpose of this will emerge in my discussion of Henry Bates's work, later in this chapter.

Enchanting as this case may seem, there is nothing magical about ultraviolet light; it is just another colour in the rainbow. But again, it does vary in content depending on the time of day – there is little ultraviolet present at dawn and dusk. It is the colour that transmits least well through the atmosphere, and can be almost completely absent under forest canopies, where light bounces around like a pinball and is absorbed by the leaves. Now it is time to consider light as a creator of niches – 'ways of life' for animals.

West Indian *Anolis* lizards inhabit forested areas. Different species reveal different colours, and it is easy to assume that their colours simply attract their own species within a busy environment. Their environment is busy – the forest contains a variety of microhabitats, constructed by the physical nature of the plant life – but the *Anolis* lizards are not all spread throughout the entire forest. They do all occupy the same forest, but they divide up the height or profile of the plant life into microenvironments based on light conditions, including ultraviolet content. And the colours of each species are adapted exactly to the light of their specific microenvironment. So in each microenvironment, one type of colouration will be most adaptive, and the owners of that colouration will be the most successful there. In their correct microenvironment they can attract mates and defend territory most efficiently, allowing them to devote more time and energy to other activities. In this case, light is the foremost stimulus. The *Anolis* lizards have adapted to light most significantly, and other selective pressures secondarily. Adaptation to light is necessary for survival. A similar story could be told for many other animals, including birds and fishes.

A more unusual form of adapting to light is found where animals take their colour directly from their environment, without drawing on their body chemistry. The pink colour of flamingos derives from the carotenoid pigments in their crustacean food. And in a case of camouflage, flatworms parasitic on marlin take up pigment from the marlin's skin below them to match their backgrounds and effectively disappear. But other animals, including the cuttlefish and chameleon in some situations, use chromatophores to gain camouflage. The skin may be equipped with sensors that detect the colour and brightness of the animal's immediate background. This is possibly the ultimate in adaptation to light. A disguise from predators can be conjured up in any environment, and then warning or mating colours can be flashed when appropriate. But when chromatophores are not a possibility, the balance between direct and indirect protection can, throughout evolution, tip one way . . . and then another.

The Victorian naturalist Henry Bates spent the years between 1849 and 1860 wandering the Amazonian rainforests. After collecting ninety-four species of butterfly, he published an article. That article has generated heated discussion ever since.

Bates grouped together his butterflies based on their colouration, as did every collector of the day. Some nice relationships emerged – the butterflies with similar colour patterns could be placed neatly into apparently related groups. But then Bates discovered some conflicting evidence – the shapes of the butterflies' bodies told a different story. Wings apart, the shapes of the body and limbs varied considerably within a supposedly related group. In fact new groups could be formed based on body and limb shapes alone, groups very different from those based on colour. So why did unrelated butterflies share the same colouration? Was this simply a 'wonder of nature', according to pre-Victorian philosophy?

Darwin and Wallace had demonstrated that wonders of nature do not exist, and Bates shared their views. He delved deeper into his dilemma of contradiction and noted that the most brightly coloured butterflies also flew the slowest, making them the easiest of prey for birds. Bates concluded, however, from the lack of evidence from discarded wings, that birds avoided them. From this he assumed these defenceless butterflies were unpalatable. Then followed an assumption which had serious repercussions – that birds understood the butterflies were distasteful based on their colouration.

Now Bates could explain his relationship dilemma. It was the shape of the body and limbs of butterflies that marked their true evolutionary relationships. While many within a genuinely related group did possess similar colours, some had departed from the norm with a purpose – that of enjoying a greater chance of survival. First, a butterfly group that has not evolved distasteful chemicals may evolve camouflage colouration, like the peppered moth. But if the camouflage code can be cracked under certain circumstances, then another evolutionary option is to pretend to be unappetising – to copy the colours of those armed with distasteful chemicals. This behavioural and evolutionary strategy is known as mimicry.

The precise mechanism of mimicry and colouration that warns of indiscernible defences is a subject in its own right. Especially interesting is how predatory species and individuals learn to interpret visual warning codes without wiping out the potential prey species in the process. John Maynard Smith is particularly well known for untangling this academic web in the twentieth century. But for the purposes of this book it is enough to know that mimicry *does* work. After all, it exists in abundance, and that's the real proof.

In his statement on colour, Darwin used the words, '*whenever* colour has been modified for some special purpose . . .'. The use of the word 'whenever' is interesting here. Does this mean that colour can sometimes be incidental in that the colour effect has no purpose?

Black sharks may be red herrings as far as colour is concerned, where their colour provides a warning only to biologists studying adaptations to light. Kanoeohe Bay in Hawaii is a nursery for the scalloped hammerhead shark. The shark pups prefer the safety of the sea floor, even though this reaches a depth of between only 1 and 15 metres within the bay. At the deepest part of the bay, the pups are almost white in colour, but at the shallowest parts they are black. The sea floor, on the other hand, is consistently white. So do the pups require camouflage from a predator or prey only in the deeper water? Or are they making a statement with their colour in the shallows? In this case, the answer to both questions is 'no'. Sometimes colour has no visual function and is said to be incidental. An example of this is the blue we see of our veins.

Pigments can serve a function other than providing a visual effect. The black or brown pigment melanin can increase the strength properties of a structure, such as a beetle's exoskeleton, or it can provide protection from the sun's ultraviolet rays. For many animals, ultraviolet light can cause tissue damage. Just as we tan in the sun, in very shallow water the hammerhead sharks do the same. At a depth of 1 metre, the ultraviolet content of the water is six hundred times greater than at 15 metres. So in the shallowest waters the hammerhead pups were gathering a layer of melanin in their skin. Melanin not only absorbs the harmful ultraviolet light but also other wavelengths or colours. Consequently, no light is reflected and the shallow-water sharks appear as a colour void. Black, that is.

This function for pigments has no place in the literature of colour, and rightly does not appear in the *Origin*. Darwin did, nonetheless, omit a function in his bold statement on colour – the 'wolf in sheep's clothing' function. We have seen examples of camouflage for indirect protection, against one's enemies. But camouflage colouration can also be employed to conceal oneself from one's prey.

I had my first active encounter with pigments while snorkelling in Greece. Although there were no coral reefs, the water was remarkably clear, blue and inviting. In the shallow water were large, brown rocks distributed randomly on the white sea floor. I noticed the rear end of a bright yellow fish emerging from the gap between two rocks and dived down to take a closer look. At first I saw nothing unusual, although I did wonder why the fish did not swim away in my presence. It appeared, through my naive eyes, to be almost jammed between the rocks, so I reached out to help it. Just as I touched its tail, something moved. Not the fish, but the rock. Part of the rock 'changed' slightly, and, on closer inspection, that part turned out to be an eye. The rock was a rock-plus-moray eel. A large brown moray eel, camouflaged perfectly against the rock it had wrapped itself around, was grasping the yellow fish head first in its gaping jaws. I was young and, since the yellow fish was about the same size as my head, I felt it was time to leave the water. Later I learnt that generally in shallow seas fish must beware all rocks . . . and stones.

This colour, like that of the Great Barrier Reef and the Amazonian angelfish, was apparent in very shallow water. It was within a few metres of the surface. And this is why we see the full spectrum of colour on the reef. If the reef were deeper, its spectrum would be limited considerably.

Monet's paintings taught us that on land the colour of sunlight changes with the position of the sun in the sky. This also happens in the sea, but there is another factor affecting the sun's spectrum under water – depth. As sunlight travels through water, it becomes absorbed and eventually disappears. But it does reach a kilometre in depth, at a level that can still be detected, albeit extremely faintly. However, as we plunge further down into the sea, different wavelengths or colours are absorbed at different rates. Red, ultraviolet and violet are the first to fade away, and at 200 metres sunlight is exclusively blue. But regardless of depth, blue transmits best through seawater, even in the shallows. This effect is quite noticeable. Diving beyond about 10 metres, the world appears blue-green. And, as expected, animals are adapted only to the colours left in their specific environment.

Below 200 metres, many animals are red. The light here is blue, and only blue. The lack of red light means that red pigments have no chance to reflect. Instead they absorb the blue light and so appear invisible. Red is a good camouflage colour in the deep.

A problem faced in mid-water is how to appear camouflaged from both above and below. From below, a fish is viewed against a light background – the sky. From above it is viewed against the darkness of the deep. The answer is to have a dark upper surface and a pale lower body. This strategy of 'countershading' is common under water, so again it must work. The marlin is a fish that appears conspicuously coloured when out of water. But put it in the water and its hues and patterns take on the roles of countershading and disruptive colouration, and the fish disappears from sight. The marlin is a huge fish, yet it can swim in front of your very eyes without your knowing it. It may be camouflaged either against predators (sharks), its own prey (smaller fish), or both, and the camouflage is so important to the survival of the fish that even the parasites on its skin have to maintain the camouflage. Abigail Ingram, a postgraduate student at Oxford University, has found that the sea lice of marlin possess chromatophores, so they can maintain the fishes' camouflage whether they occur on dark or light areas of skin. This is a different strategy from that of the flatworm parasites of marlin that steal the marlin's pigments, but it has the same result. If the marlin dies, so do the parasites. And then there are sucker fish, which clean the marlin of its parasites, to consider. So marlin parasites must appear camouflaged to these fish too. Consequently light is a selection pressure acting on the marlin *and* its parasites.

Victoria Welch, another postgraduate student at Oxford, has been tackling another form of camouflage. Countershading is a possibility for fish because generally they remain horizontal. But some animals vary their orientation. Jellyfish often roll around in the water and effectively have no upper and lower surface. They lack the sophistication in hardware and software to handle chromatophores and are often left with only one option to help them blend into their background – transparency.

Throughout their evolution, many jellyfish have bypassed the road to colour matching. Instead, these jellies blend into their backgrounds using the background light itself – it shines right through them. But this solution is not that simple. Jellyfish often can maintain transparent innards – that is not their biggest problem. Victoria Welch is considering some less obvious stumbling blocks – polarisation and surface reflections.

Predatory fish can detect light that is polarised. Consequently, selective pressures act on the jellyfish to avoid becoming a polarisation filter. Light of all polarisations must pass through the jellyfish, and not just some polarisations. If this demand is not met, the jellyfish will match its background light in terms of colour, but not polarisation, and so it will not be completely invisible.

And then there are the surface reflections. We see a reflection of ourselves in glass windows – the effect of any completely smooth surface at the microscopic scale. But jellyfish must not act like glass and reflect only some light from their very outer surfaces. Indeed, jellyfish may have surfaces that reduce reflection considerably, in which case the potential problem is solved.

Light is certainly a major force in governing the behaviour of animals today. And for life to have reached this point, light must have been a considerable factor of evolution in the past. Such thoughts will be pursued later in this book; they will form another piece of the Cambrian puzzle. But the current subject, light in environments today, will emerge as perhaps the most important clue of all in solving this enigma, although a less obvious piece of the puzzle at this stage. Certainly this is a subject into which we should delve as deeply as space will allow. But we should understand what we are *really* dealing with here, and not lose sight of this throughout the rest of this book. I refer to an animal's *complete* visual appearance.

The officer's hat, or the relevance of size and shape

In my earlier description of an eighteenth-century soldier's uniform, I touched on another point relevant to nature – size and perceived appearance. The balance between visual camouflage and conspicuousness is not influenced by colour alone. Colour in nature is not the sole component of an animal's visual appearance. Size, shape and movement relay considerable information, too.

Just as soldiers wore exaggerated headwear to appear larger, and consequently a greater threat to the enemy, so the puffer fish inflates itself when danger approaches. And when a toad encounters a snake, the toad instinctively stands on fully outstretched legs and inflates its body to some three times its normal size. Now the snake registers a different image – from one that looked like an easy meal to one that has become a possible aggressor. Suddenly the snake is less likely to attack, its judgement based purely on the visual appearance of the other animal. And in all environments with light, visual appearances as a whole influence interactions and relationships between species.

More obviously, the shape of an animal is an important component of camouflage and mimicry. The stick and leaf insects, and weedy sea dragons, must possess the colours *and* shapes of sticks, leaves and seaweeds respectively. The movement of these animals is just as vital. The praying mantis that mimics leaves must sway in the wind just as the leaves around it does.

These are physical and behavioural adaptations to light. Light not only affects the colour of an animal but its whole form and behaviour. Remember that if an animal is not adapted to its light environment it will not survive. Now we can see that this rule calls for great responses throughout the evolution of a species. It is not enough for a lioness to have beige pigments that allow it to blend into the surrounding grass. The lioness cannot evolve the contours of its environment so it must possess another weapon to enable it to catch food – it must be capable of keeping a low profile, not unlike a military sniper. This again is an adaptation to light. But then the lioness's prey is itself adapted to light too. Wildebeest often graze in circles, facing out from the centre. They are looking out for the lioness, and now collectively they can scan the entire plain. Their circle is also a behavioural adaptation to light.

To successfully achieve camouflage, even shadows must be considered. A green beetle on a green leaf is not camouflaged if it casts a shadow. But again evolution has responded so as to make life difficult for predators. Many beetles living on leaves are hemispherical in shape. This is a physical adaptation to light. A sphere will always cast a shadow, but from most positions a hemisphere will not. It is important to consider the tremendous evolutionary cost of yielding such a change to the standard beetle design. Not only is the body affected, but also the legs and wings, and any walking and flying that subsequently takes place. Light must really be a powerful stimulus.

Shape and behaviour are important components of conspicuousness, too. Bees perform dances that carry in them directions leading to nectar. Their wiggles and pirouettes are all adaptations to light – they are visual signals. More familiar, the peacock displays spectacular colours to the comparatively drab peahen. Of relevance here are the eyespots, apparent at the tip of each tail feather. The number of eyespots are 'counted' by the female, and as far as she is concerned the more the merrier. The static peacock may possess a hundred eyespots, but during courtship the peacock is not static. He shakes his tail feathers.

As a comparison, hold up a pen and shake it rapidly from side to side. The single pen will appear as two, one at each extremity of the movement. The same thing happens to a peacock's tail feather. When shaken, two eyespots emerge from the single feather. Now consider the complete peacock. When his eyespots are under scrutiny, he shakes them, so that his hundred eyespots become two hundred, indicating a much fitter individual. No peahen would be content with a mere hundred eyespots. And again, this shaking behaviour is part of an adaptation to light.

Eyespots are common in the animal kingdom. Often they perform the 'officer's hat' role – that of making their host look bigger. A butterfly with eyespots at the edges of its wings appears as one large head to some potential predators, in which case the whole animal becomes conceptually much larger. But not all predators are so easily fooled. And eyespots can have further drawbacks.

Pictures of butterflies in reference books tend to show their wings from above. But sometimes potential predators or partners approach the butterfly from an angle. Then, the eyespots become eggspots – the circles appear elongated.

In 1533, the German artist Hans Holbein the Younger painted one of the first portraits showing a full-length, life-size person. *The Ambassadors* features two men and at first everything appears quite normal . . . until the observer notices a strange, elongated object at their feet. The painting was originally hung at the top of a great staircase, where it could be seen either straight on or from an oblique angle. As described, the view is from the front, but from an oblique angle at certain points on the stairs the two men become blurred and eternal death takes over. The men no longer are perceived as bodies, and the mysterious object reveals itself as a distinct human skull (see Plate 10).

During the courtship of some butterflies, the male views the female's wings from an oblique angle. Any female patterns that serve to attract the male must, therefore, emerge from an oblique view. The results of evolution may not be those that first meet the human eye. Up to this point I have discussed pigments, but the effect of the viewing angle will become even more significant when structural colours are considered.

Now we can begin to understand why the animals we see are the colours they are, and also that there is a degree of sophistication in the system. But in fact we are merely at the base of the scale of sophistication where nature's colours are concerned. There is another way that animals can appear coloured other than by employing pigments. Ironically, the brightest colours in nature result from purely transparent materials.

Structural colours

The Romans were highly skilled in the art of glass making. We know this because in Roman burials artefacts were often placed in the coffins of the dead, and among those artefacts were pieces of glassware. Much of that glassware has survived and been recovered completely intact.

One Roman glass plate in particular caught my eye because, not only was it complete and undamaged, it bore an iridescent glaze. Reds, yellows, greens and blues – in fact a complete spectrum – were shining from the surface of this specimen. The colours possessed a metallic appearance. They were much brighter, or more noticeable, than a plate painted with pigments. Interestingly, the colour observed appeared to change as I moved around the plate, an effect also seen to take place on the otherwise transparent wings of a housefly. But what do the Roman plate and an ordinary housefly have in common?

The glass plate had an extremely thin, fragile coating, one that could easily be rubbed off by hand, leaving a plain, transparent glass plate. The coating was obviously responsible for the metallic-like colouration. It is quite literally a 'thin film', which may seem a fairly broad description. But in the world of optics this term means much more than that. And again it was Newton (or possibly Robert Hooke) who first realised that thin films also occur in nature, when he deciphered the cause of the peacock's iridescence. Newton, in fact, gained his ideas from thin flakes of glass.

The following few pages are devoted to describing the commonest structural colours in nature. The mechanism behind these metallic-like colours can be interesting if only because they explain the paradox of their transparent foundations. But more incisively, it will demonstrate that physical structures really can cause colour – an understanding which will turn out to be invaluable later in the book.

Unlike chemical pigments, physical structures are preserved in the pickled collections of natural history museums. So one can study their cause and diversity without requiring access to living specimens; that can be very useful. But, as revealed in the previous chapter, physical structures have been preserved in animals other than those living today. In order to help paint a more informative picture, it is worth persevering with a short lesson in optics.

Simply put, a thin film is a thin layer of material. It only has upper and lower surfaces. In terms of its effect on light, the material acts as a different medium to air. Descartes demonstrated that light reflects from the outer and inner surfaces of a water droplet, and Fermat explained this in that light travels at different speeds in different media. A thin

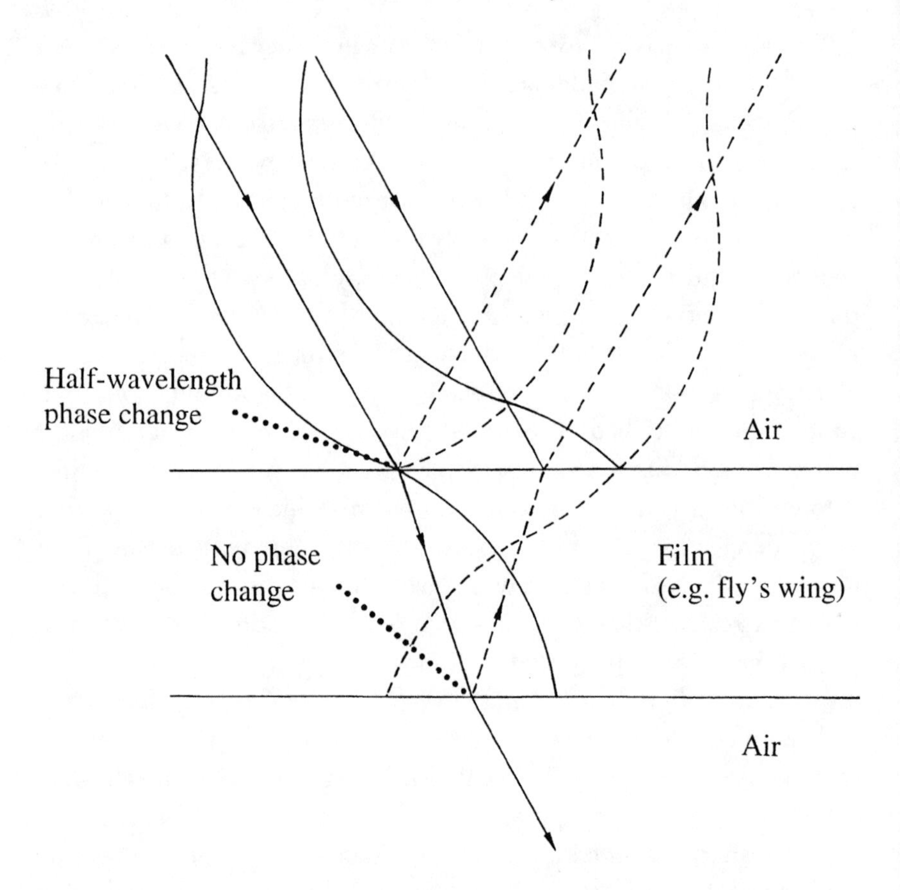

Figure 3.2 Light rays affected by a thin film, such as a fly's wing, in air. The film is shown in cross-section; the light-ray paths and wave profiles are illustrated as solid lines (incoming light) and dashed lines (reflected light).

film of transparent material also acts like a water droplet – light reflects from its upper and lower surfaces. Maybe about 4 per cent of the rays in the original beam reflect from each surface of the thin film, and 92 per cent pass through the film.

When the reflected rays are out of phase, they cancel each other out like the ripples caused by Leonardo's stones. In this case, a beam does not exist. But when the wave profiles superimpose, they are said to be in phase and a light beam does exist. In which case there is a 4 + 4 per cent, or 8 per cent, reflection. This may seem trivial, but remember pigments appear as less than 1 per cent reflection because their reflection covers a hemisphere, and we see only a tiny segment of that hemisphere. The 8 per cent reflection from the thin film, on the other hand, travels in a single direction. So if we view the film in that direction, we see the full 8 per cent. Consequently the thin film appears much brighter than a pigmented material under the same illumination (although not *that* much brighter because the eye is a logarithmic detector). This situation exists when the thickness of the thin film is about a quarter of the wavelength of light.

By introducing a change in medium, Newton's prism caused different wavelengths, or colours in white light, to travel in different directions. If we apply this concept to our thin film model, we also get different colours reflected in different directions. In one direction the reflected waves for only one colour can be in phase, and the others will be out of phase and will not oscillate any more in this direction. So the colour will appear different as we view the film from different directions. And that is the effect we get from the Roman plate and the housefly's wing. Soap bubbles and oil slicks are also thin films.

I have attributed the 4 per cent surface reflection of our thin film to a change in media. I didn't mention, however, that different media cause different reflectivities. The 4 per cent condition occurs between glass and air, but this is reduced when the glass is placed in water. This has happy consequences for the transparent jellyfish, with a skin of similar optical properties to glass – reflections from their body in water will be slight, much less than in air. And it is the floating relatives of the jellyfish, like the Portuguese man-of-war, that suffer most from the pitfalls of reflectivity. Parts of their bodies are naturally exposed to air.

Previously I have explained the workings of the jellyfishes' dream – chromatophores, colour cells that cause the changes in skin hues originally thought to result from chemical reactions. Although an erroneous explanation, the chemical reaction scenario was a theoretical possibility. And there is yet another possibility.

I once attended a public lecture on liquid crystals. Liquid crystals contain helical molecules, slotting together like a row of tiny springs. And I do mean tiny – side on, each complete turn of the coil measures only half a wavelength of light. That is, half the wavelength of the light that is reflected from the structure – liquid crystals can appear strongly coloured. We can buy toys or thermometers containing liquid crystals – those that change colour with a change of temperature, as when they are touched.

The colour of a liquid crystal derives from light waves that reflect from each half turn of the helical molecules. This can be best understood by considering the whole structure as a stack of thin layers, where the materials of alternate layers have different optical properties.

Each half turn of the helical molecules covers a quarter of a wavelength in distance or 'thickness', and is now equivalent to a single thin

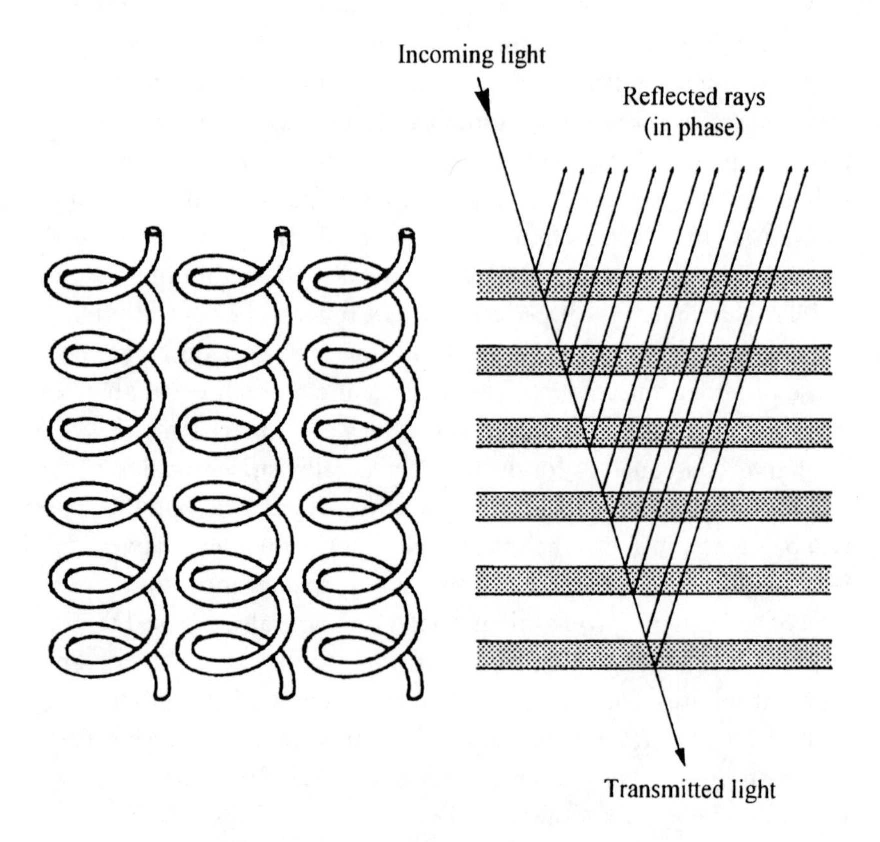

Figure 3.3 A cross-section through a liquid crystal (left), showing individual helical molecules, and its approximation as a stack of thin layers and effect on light (right). Reflected light rays are in phase when the layers are approximately a quarter of their wavelength in thickness.

film. So the whole molecule from top to bottom approximates many thin films piled up. As a stack of thin films, the liquid crystal molecules can reflect a greater portion of the light. The 92 per cent of light that passes through a single layer meets with another layer, and another 4 per cent reflects. Eventually, with enough turns of the helical molecules, all the light will be reflected and none will pass through the system. Now we have 100 per cent reflection – the brightest possible effect, if you are looking in the right direction.

During the question time that followed the liquid crystal lecture, a member of the audience enquired, 'Does the chameleon change its colours using liquid crystals?' A very nice thought. In fact, it was so nice that the lecturer responded with an enthusiastic, 'Yes, you could be right!' The lecturer was a chemist and could be forgiven for not knowing the real cause of the chameleon's guise, but the question demonstrated an excellent understanding of the whole lecture. The lecturer had succeeded in getting her message across and was clearly so delighted that a negative response would have seemed inappropriate. Liquid crystals, nonetheless, can be found in the literature on animal colours.

Down House, in south London, was the home of Charles Darwin. His original microscope, desk and bookshelves are preserved intact, along with some of his specimens. Although barnacles were Darwin's chosen group, beetles seem to have followed him everywhere. And within Down House, as within Oxford's Huxley Room, shine the metallic effects of liquid crystals, naturally embedded with the exoskeletons of Darwin's beetle collections.

Beetles are often so spectacularly coloured that they end up adorning the costumes of New Guinea tribal chiefs or are sold as the earrings in jewellery shops. Their metallic colours appear to be more diverse in the tropics, which is not so surprising because here the sun's rays are stronger. The lack of cloud cover results in up to twice the luminance of temperate regions. So the selective pressures for bright colouration are stronger in the tropics, and evolution has responded accordingly.

Although liquid crystals approximate multilayer reflectors, true multilayers also exist in beetles, particularly tropical species. Observe the broken wing case of a Thai flea beetle in an electron microscope and a clearly defined stack of thin layers becomes evident. But then spongy structures can sometimes be found in the wing cases of beetles, which, like the liquid crystal, work in a similar way to the true multilayer reflectors.

H. E. Hinton was an entomologist at Bristol University in England with a strong interest in colour. In 1971 he was collecting insects in Venezuela. But his most exciting find came not from his trap samples, but while he was filling his car with petrol. A male Hercules beetle, the second largest insect in the world, flew into the strip lighting of the petrol station and fell to the ground. This would have been quite a sight – in flight, this beetle looks like an armoured bird. Hinton got to the stunned beetle first and quickly placed it inside a sock from his luggage. The beetle's horns became stuck in the sock, which turned out to be the perfect prison. Hinton, however, was more than curious about the specimen. 'At odd moments I used to take it out and play with it,' he admitted. But of more interest to science, he added, 'In due course I became aware that its elytra [wing cases] would change to greenish yellow and back again to black.'

The wing covers of the Hercules beetle contain a spongy layer above a black pigmented layer. The holes in the sponge act as alternate layers of a multilayer reflector, and this can account for the greenish yellow colour Hinton observed. But what about the intermittent black colouration?

The above multilayer condition is satisfied when the holes in the sponge are filled with air. In such a case, light effectively recognises a difference in media and the thin layered effect emerges. But that effect disappears when the holes are filled with water, a medium with optical properties similar to that of the beetle's wing cases in this instance. Now, light recognises no boundaries as it passes through the spongy structure, and is stopped in its tracks only by the black pigment.

Hinton's beetle was observed under different conditions of humidity. Under high humidity levels, the spongy layer of the wing covers filled with water and they appeared black, from the pigment. Under low humidity levels, the air spaces were restored and the yellow and green wavelengths were reflected before they reached the underlying black pigment. The physical structure, and consequently the colour, was altering. So it's not surprising that liquid crystals and chameleons were mentioned in the same sentence in the lecture theatre. But, like the chromatophores of chameleons, do structural colours have a biological function?

Because of their behavioural effects, structural colours are easier to define than pigments. They are the brightest colours found in nature, and their visual effect must always have a function – where they occur on parts visible in the environment, that is. Structural colours do derive from physical structures, so potentially they may have another function. A broken mammoth tusk, for instance, reveals a stack of concertinaed layers internally, which lend strength to the whole tusk. The layers in this case, however, are much thicker than the size of light waves, and they do not cause colour. So change the dimensions of a reflector's structure, and the colour, but not the strength property, can disappear. This change may be slight. Consequently a minor mutation is all that's needed to put an end to a structural colour. And considering its powerful visual effect, selective pressures acting on a redundant structural colour would be strong. Redundant structural colours do occur in nature, but only on parts not visible in the natural environment. Many shells opt for an alternative to changing their internal layer thicknesses. They have shiny, structurally coloured internal surfaces, but this visual effect is masked from the mollusc's environment by an outer layer of absorbing pigment. In Darwin's statement at the head of this chapter, the phrase 'whenever colour has been modified for some special purpose' refers only to pigments. I suggest that structural colours entering the environment are always functional. They can't afford not to be.

Unfortunately, lack of space in this chapter precludes mention of some fascinating alternative mechanisms for producing structural colours, although some of these will appear in subsequent chapters. I must also omit detailed reference to the large glass cabinet to be found in Down House, stuffed with hummingbirds and birds of paradise with magnificent structural colours. Also, I have stopped before things start to get really complicated. This is where biology becomes a minor subject and complex electromagnetic scattering theory, deep within optical physics, takes central stage. The purpose of this chapter was not to explain the complete workings of natural colours, rather to hint at the range and sophistication of colour in the natural world.

More specifically, in this chapter I aimed to generate thought about how animals have adapted to light in general. This, as I have suggested, involves not only colour, but also shape and behaviour. Evolution has resulted in the refined adaptation to sunlight wherever we look. I have provided examples; it is now down to the reader to look around and complete the picture. If we can fully appreciate this great adaptation, we will have discovered a major clue to help us understand the Cambrian enigma. Thoughts assembled in this chapter will become moulded into something firmer as this book progresses, and eventually all will become clear. Crystal clear.

Light certainly is a force to be reckoned with today . . . where sunlight exists, that is.

4

When Darkness Descends

Blessed is your rising in the horizon of heaven, living Sun, you who were first at the beginning of things. Your rays embrace the lands to the limits of all that you have made

Hymn of Akhenaten, pharaoh of Egypt (1,000 BC)

. . . Well, almost to the limits.

In the second half of the eighteenth century, before the declaration of evolution, the Reverend Gilbert White wrote many letters to Thomas Pennant and Daines Barrington, acquaintances who shared his interest in the natural history of Britain. White lived in the Hampshire village of Selborne, and used the wildlife of his parish to encourage the zoological curiosity of his fellows. In 1788, more than a hundred of his letters were gathered into a single volume. *The Natural History of Selborne* became the fourth most published book in the English language.

White, Pennant and Barrington described the wildlife of Selborne, and some of the nature encountered during their expeditions around Europe, as they saw it. They painted a vivid picture, one in existence only under daylight. But did they acknowledge life at night? And did Darwin observe the fields and woodland surrounding Down House as the sun went down? The answers are 'no' and 'no' again. The previous chapter begs the question 'What about nocturnal animals?' Well, I had a reason for overlooking this subject. Night-time on terrestrial Earth is a grey area. It is neither bright nor completely dark.

Darwin, faced with a mountain to climb in any case, ventured only into the world he saw with clarity. Humans have adapted to the visual world of daytime. But a letter from Thomas Pennant to Gilbert White indicates that there also exists a visual world at night. During a tour of Scotland, Pennant noted his sighting of an eagle owl.

I once spotted an eagle owl in the heart of England. Driving home in the dark, my headlights picked out the sign for my home village. All seemed perfectly normal, except that an eagle owl was perched on the sign. Wait a minute. An eagle owl, in England? I must have been mad or drunk. But I knew I hadn't been drinking. Maybe the eagle owl, over 2 feet tall, was a figment of my imagination. I was unaware of Thomas Pennant's sighting at the time, but I knew my owls. And eagle owls do not live in Britain.

I decided to forget about my apparition . . . until I turned on the radio the following morning. Concluding the regional news was a story about an Egyptian eagle owl – one that had escaped from the local wildlife park. Suddenly I chose to recall my apparition of the previous evening. And the most memorable part of that bird was its eyes – its *huge* eyes.

What Thomas Pennant saw in the eighteenth century is no longer relevant today. Although they once lived in Britain, eagle owls reside elsewhere now. But where they do exist, they are active at night. And to catch their prey they use sound . . . and light.

In the previous chapter we learnt that through larger eyes pigments would appear brighter, because big eyes sample a larger segment of the pigments' multidirectional reflectance. At night the Earth is lit by moonlight – the sun's rays reflected from the moon. Humans cannot efficiently detect these rays and often fall short of the visual frontier at night.

Now Darwin's exclusions and the eagle owl's eyes become interesting. What Darwin could not see, the eagle owl can. The theme of this chapter is darkness, and what happens to wildlife that is deprived of light. But on a journey into total darkness it is worth adjusting our eyes via intermediate cases, beginning with the first step.

Night-time on land

Without the aid of night-vision equipment, it is not surprising that Victorian and earlier naturalists concentrated their efforts on daytime. But while they gazed into their perceived darkness, nocturnal rodents scurried in front of their eyes, and owls were watching them.

Mammals were never going to be champions of camouflaged shapes. Their highly sophisticated machinery, particularly their warm blood, calls for a generous volume compared to surface area – they must be roundish. Still, they try their best to be camouflaged, as with the lioness hiding itself in the grass. They have succeeded with background-matching colours, but sometimes that is not enough, in which case they are compelled to evolve in darkness.

It is interesting that on land the same physical environment exists at night as it does during the day. Trees and rocks continue to provide nooks and crannies . . . but no longer areas of brightness and shade. And the evolutionary outcome? There are considerably fewer species active at night compared with the day. There really are fewer niches – 'ways of life' – available at night.

The reduction in niches caused by the lack of light is central to this outcome. And then comes the secondary factor – feeding. Ripples travel down the whole food pyramid. Fewer niches lead to fewer species near the base of the pyramid. This in turn narrows the whole pyramid, where at the top there are fewer predators. But the night-time pyramid occupies the same physical space as that of the day-time pyramid. So the food web becomes stretched and offers less opportunity for tangling, or for evolution to cross lines. Evolution maintains a comparably low diversity at night.

Heat is partly responsible for this. It is warmer during the day than at night, and many animals are adapted to warmth. But animals from most phyla *can* be adapted to the cold. This is not an evolutionary impossibility. So we can consider at least part of the day–night biodiversity difference as evidence towards the power of light as a stimulus affecting life on Earth. Begin to remove this stimulus and evolution becomes much less complicated. I say '*begin*' because night-time on land is only a step towards total darkness.

At night, other senses are employed. But this is where the big difference between light and the other major stimuli becomes clearly evident. I refer to the difference in presence. Light strikes the Earth and oozes through the canopies of trees, between rocks and blades of grass, and into the waters – it cannot be avoided. Light infiltrates an environment whereas the other major stimuli do not. This explains why owls, equipped with extremely sensitive hearing and the potential to further develop other senses, do not relinquish their use of light. In fact vision has evolved further in owls. A mouse that has detected the flight of an owl may freeze and become inaudible – the equivalent of invisible to light. But where invisibility demands great evolutionary effort, inaudibility requires only temporary stillness.

Up to this point I have considered the major senses – senses that are common in nature. These are smell and taste (which are quite similar), sight, hearing and touch. But at night, one of the minor stimuli becomes important. This stimulus carries the advantage of light in being unavoidable. As described in Chapter 3, bats hunt using radar.

Radar is a minor stimulus/sense as a result of requiring considerable evolutionary expense and chemical and mechanical effort just to infuse the stimulus into an environment in the first place. Light, on the other hand, is a pre-infused stimulus. Only *then*, when radar has been launched into the air, can its detection be compared to vision. And even so, a bat's radar invites little evolutionary change in the animals not directly affected by this stimulus. Light, on the other hand, affects everything in the environment where it exists.

The owl is completely unaffected by the bat hunting moths around it. During the daytime, however, apparently isolated predator–prey relationships begin to interact with each other. The food web and animal behaviour become increasingly complex. So in addition to the direct reduction in niches at night, through the degeneration of light and shade partitions for instance, evolution is stimulated much less at night. Again, in this chapter I place emphasis on the predator–prey scenario because the first rule of survival is to avoid becoming a meal. So this interaction is as important as it gets.

On land, the transition from light to almost dark happens quickly, during sunset or at dusk. So few animals on land are adapted to anything other than light or almost dark conditions. But in the sea there is another transition from light to dark – a transition in space. Marine animals can be compared from different depth ranges, living under different light levels.

The biggest clues towards solving the Cambrian enigma from night-time on land are the reductions in both biodiversity and complexity of behaviour that accompany a reduction in light. We will develop this understanding throughout this chapter, but further clues can be found in the deep sea, where evolution within a tiny branch of the animal tree can be tracked through time.

The deep sea

The Scavengers of East Australian Seas, or 'SEAS', expedition was established to scientifically document the entire community of scavenging crustaceans – the group of arthropods that include the crabs, shrimps and lobsters – along the east coast of Australia. Before 1990, traps were set for these animals, but these were poorly designed and caught only individuals bigger than a few millimetres. In fact a twelfth-century fish/crayfish trap was recovered from the River Thames where it passes the Tower of London, and its design turned out to be superior to twentieth-century traps. Its overall form was that of a wickerwork cone, with a funnel-like entrance. Beyond the entrance lay an additional but narrower funnel-like entrance, creating two chambers inside the cone that could hold catches of different sizes. The victims would have been lured into the cone by bait in the smaller chamber. The whole trap was weighed down on the river-bed by two large flints, and connected to the surface by a rope.

Not only had scientific scavenger traps fallen below twelfth-century standards, but they had been set sporadically – on a random basis within small areas, and without the bigger picture in mind. Jim Lowry had been thinking about this lax approach for some time, and decided *he* would paint the bigger picture, and in turn lay the foundations for scavenging crustacean conservation.

Scavenging crustacean communities are exceptionally important

Figure 4.1 A twelfth-century fishing trap recovered from the River Thames.

because they clean the sea floor of dead organic matter such as fish carcasses, which would otherwise consume valuable oxygen in the water as they decayed. And throughout the course of a normal day there is quite a fall of bodies to the sea floor. Also scavengers are a noteworthy part of the marine food web – they in turn provide food for other inhabitants of the sea, and complete the cycle of organic nutrients.

Jim Lowry had moved from Virginia in the USA to the Australian Museum in Sydney via a lengthy spell in New Zealand. He chose his back garden as a study site – the east Australian coast, in fact, and no small undertaking.

Jim Lowry lives on a small island within a marine inlet to the north of Sydney. He travels to work by motorboat and motorbike. His bike is a beautiful, black and chrome 750cc machine. His boat is rather less impressive, but is affectionately known as 'The Flying Scud'. Scud is the American slang, though not quite a household name, for amphipod – a type of crustacean. Amphipods are commonly encountered on beaches, near rock pools, in the form of 'beach fleas'. Often they have shrimp-like bodies that are flattened from side to side. Jim Lowry studies amphipods. He produces (along with his co-worker, Helen Stoddart) some of the finest taxonomic work to be found anywhere.

Taxonomy is arguably the oldest scientific profession. It involves documenting and describing new (to science) species using consistent methods, and is one of the most essential of all scientific disciplines. Scientific classification began with the Swedish botanist Carl Linnaeus in the eighteenth century. We still use his system, but Darwin and Wallace's theory of evolution has allowed scientists to see diversity as the result of a dynamic process rather than a static picture. Considering the extinction rate induced by humans, and that only about 10 per cent of the Earth's species have so far been described, we should really be in a hurry to get on with taxonomy. Taxonomy is also important from an evolutionary perspective. We must describe and collect nucleic acids from the species alive today in order to perform evolutionary and genetic diversity analyses. Better to collect DNA from species while they are alive rather than extinct. Remember the drama of collecting ancient DNA from just a single extinct species such as the mammoth? Unfortunately we have been a little slow off the blocks, to say the least. Today species are disappearing faster than they are being described.

Jim Lowry's interest in scavengers stems from the amphipod connection – amphipods are among the chief scavengers. The other principal scavenging group was thought to be isopods. Isopods are also shrimp-like animals but their bodies are typically flattened from top to bottom, rather than side to side. Isopods include woodlice – the only members of the group with any notoriety, although probably bad examples since most isopods are marine.

Jim Lowry designed a scavenger trap not too far removed from the twelfth-century model. Plastic drainpipes were sectioned into short tubes to form the frame of the trap and to provide a robust structure; his traps were destined for deeper waters. Plastic funnels were cut accordingly to provide two different apertures, and they were glued into the 'drainpipe' tubes to form the two chambers. A mesh was fixed at the end of each tube to allow water to flow through the trap rather than sweep it away. The size of the mesh was important – holes half a millimetre in size were selected, allowing anything smaller to escape, but anything larger to be caught.

The traps were tested near Sydney. At the Australian Museum thick rope was sectioned into 50-metre lengths. This was coiled carefully – a

Figure 4.2 A typical scavenging isopod, amphipod and ostracod (seed-shrimp).

poorly coiled rope can quickly resemble a plate of spaghetti and be just as useful – and baled on to 'The Flying Scud'. House bricks were also loaded on to the boat, along with orange plastic buoys. 'The Flying Scud' was towed out to the coast, stopping at a petrol station along the way to buy frozen pilchards. Frozen pilchards sell well in Australia, where fishing is popular and pilchards make suitable bait.

On board 'The Flying Scud', a pilchard was placed inside the smallest chamber of each trap. The traps were individually tied to two house bricks and one end of the 50-metre length of rope. The other end of this rope was tied to a buoy, and the whole apparatus was then hurled overboard to a depth of 25 metres. The length of the ropes had to be greater than the depth of the water in which they trailed, so that, as the traps rested on the seabed, the rope provided some 'give' in strong currents. Notes were made of the positions of each trap in relation to objects on the shoreline.

The following morning, 'The Flying Scud' returned to the study site to recover the traps. Finding the buoys was not so easy, and some traps were lost. But the traps retrieved were opened on board – and everyone was happy. Jim had caught his amphipods and isopods. The test run was a success, although it was clear that improvements to the protocol were necessary if the more turbulent seas off Australia were to be prospected. The marine snail community introduced one unanticipated problem, however. Sometimes they too were attracted to the smell of fish, and in a frenzied bid to dine on the pilchard, they became jammed in the entrance hole, thus spoiling the traps. News also arrived from the fisheries industry of some gigantic isopods living in deep waters off the north-east coast of Australia – and they were feeding on dead fish. All of this called for adjustments to the trap design.

Jim Lowry opened his map of the south Pacific and pinpointed his targets. Several towns were marked at different latitudes, from New Guinea in the north, traversing the eastern Australian coast to Tasmania in the south. From each town, traps would be set along a line of latitude, beginning at 50 metres deep and ending at 1,000 metres. The expedition was starting to get serious.

Behind the scenes, the SEAS project was taking shape. Jim managed to recruit several students and technicians at the Australian Museum to work on his new traps – the deadline for his first boat launch was approaching. A production line unfolded and the finished traps were piled on to a huge trailer at great speed. The new traps were all covered with metal grids to keep out the snails. And to counter the giant deep-sea isopods anticipated, the traps were placed inside much larger structures, which were actually modified lobster traps. All the equipment could be stacked, so a single trailer, albeit fully laden, was adequate.

The deeper sampling sites called for a bigger boat, and 'The Flying Scud' was retired. Commercial fishing vessels were chartered from each town, and these were equipped with a global positioning system, or GPS. This system employs satellites to locate precisely any coordinates, even at sea, and so traps could theoretically be found easily. But to fight the stronger currents in these deeper seas, and keep the traps in their original positions, anchors and heavier lead weights entered the equation. A certain amount of drift was still predicted, so the markers at the surface were upgraded too to prevent them being dragged under. Huge buoys and flags were tied on to the cage-like trailer, which was beginning to look like a travelling circus wagon. And the great bundles of rope, now up to a kilometre and a half long, only confirmed the resemblance.

The SEAS bandwagon rolled into Cairns in north-eastern Australia in 1990, and the expedition was launched. Everything went smoothly. The traps were set one afternoon, and most were collected successfully the following morning. Amphipods and isopods were recovered, and nearly all were new species. The Australian Museum jeep towed the gear to the next site and the sampling continued . . . and so on. At each port of call a different fishing vessel awaited, each equipped with a different captain and crew.

The SEAS project was a great success in that hundreds of new species were recovered during the original sampling expedition and in the repeats. Interestingly, as will become evident, the species tended to get larger as the depth increased.

The ecological results of the SEAS project are in the throes of being published. All I can say here is that they reveal, for the first time, the fate of the better known fish and other marine animals in one of the largest environments on Earth. For the first time we will understand the biology of the crustacean scavenging community, which will have all sorts of implications in fisheries practices and management. We won't be able to produce a management plan for the seas, and ultimately preserve our fisheries industry and marine biodiversity, if we don't know what's down there. The SEAS project is a wonderful success story, but it was the results of the isopod part of the research that are relevant to this chapter.

Steve Keable, a member of the SEAS team interested specifically in the isopod catches, set some traps by hand in shallow water on the New Guinea leg of the trip. He did catch isopods, but decided to cut his plans short when, surfacing from a dive one day, he spotted a local tribesman standing astride a large rock, bow in hand with an arrow strung and pointing in his direction. Steve continued with his shallow-water sampling in safer waters off Australia, and with considerable success. Faced with so many new species of shallow-water isopods, he left the deep-water species to Jim Lowry, who could not resist these amazing forms.

It was the shallow waters that revealed the greatest diversity of scavenger species. As the trap localities became deeper, the number of species caught became fewer. The total number of individuals became fewer too, but not so the total weight of the catches – the animals were getting bigger. And they were dominated by those giant isopods that fishermen had warned of, known as *Bathynomus*. *Bathynomus* was no longer a myth to the SEAS team.

The deep-sea traps were hauled to the surface by a winch. Each trap came into view in the water as it was lifted closer to the ship, at which point members of the crew leant over the hull to heave it on board. It was immediately obvious there was a living animal in the trap. Large crab-like legs began to poke out through the holes in the large outer trap, and scraping sounds were heard as sharply pointed feet crawled over the rigid plastic sides of the trap. The whole trap moved around as it lay on the deck, surrounded by the onlooking crew. Then the trap was opened.

Everybody gasped. What appeared beggared belief, best described as something out of science fiction. One is invariably taken aback by an encounter with the unknown, and here the crew were witnessing something they had never seen before – not on TV, not in books and not in aquaria. By science fiction I refer to movies about aliens or, more appropriately, those 1960s cult classics where giant tarantulas or ants chased helpless humans some ten times smaller than them.

Out of the deep had risen an isopod that looked like a woodlouse. But this creature could never be mistaken for a woodlouse – it was fifty times bigger. This was *Bathynomus*. The fishermen's legend had come to life, and giant, robust isopods were now roaming the deck. At fifty times their normal size, the jaws of a woodlouse look quite fiercesome, and their steps seem almost mechanical. Their heads, face on, look like stormtroopers from *Star Wars*, and their bodies resemble small but significant tanks, some half a metre long. *Bathynomus* indeed appears more machine than animal (see Plate 13).

It took a while for the unfamiliar to become the familiar, and the sight of a *Bathynomus* scurrying across the deck like an armoured vehicle, with jaws chomping, continued to be breathtaking. Those fortunate enough to see elephants in Africa, tigers in Nepal and bears in Canada should try adding *Bathynomus* to their list.

Something *Bathynomus* shared in common with these animals was its eyes, but *Bathynomus* lives in waters up to a kilometre deep, so what use are eyes here? Well, *some* sunlight exists, even at these depths, although only the blue component remains. And like the eagle owl, which also lives under dim light conditions, the eyes of *Bathynomus* are big. So in parts of our planet that remain too dim for us to see, but are reached by sunlight all the same, there live other animals exercising vision.

At a kilometre in depth, the sea is comparable to night-time on land in that light as a stimulus to behaviour and as a selection pressure to evolution is greatly reduced. But it is still present. This is not the completely dark scenario towards which I am aiming in this chapter, but it is a step in the right direction. Again we can learn that, where light is greatly reduced, biodiversity diminishes in unison. As the SEAS traps were set deeper, the number of species in their catches was reduced.

The deep sea is extremely interesting because there are many more amazing and unknown creatures to be discovered. New finds continue to enthral us every year. And the trend towards gigantism seems to hold, along with the low diversity levels as compared with shallower, brighter environments. Taxonomists studying sea spiders – marine members of the arthropod phylum most closely related to true spiders – also confirm that deep-sea faunas are discernible for their low species diversity while sometimes displaying an amazingly high abundance for a single species. The considerable size and weight of animals in the deep sea suggest that resources are not always limiting. But the reduction in light is a major factor in the reduction of evolution in the deep sea, implied by the depleted variety of species.

Many deep-sea animals share the 'big eye' characteristic of *Bathynomus*. Fish, squids and shrimps, to name but a few, have larger, more sensitive eyes in the deep. Evolution has continued to provide adaptations to light here, even though the light is extremely low. Light must really be a powerful stimulus. But I won't dwell any longer on the adaptation to reduced light found in animals today, partly because some animals produce their own light in the deep sea. Even where light is extremely dim, selective pressures still act on animals to be adapted to light – to see it and even produce their own, although

Bathynomus is not one of the light producers. This self-produced light, known as bioluminescence, will be described in Chapter 5. Here it may only complicate matters, although the general light field can still be described as low in the deep sea.

To get the picture of life in complete darkness we must head for caves. But before leaving the deep sea, I will return to *Bathynomus* and another lesson it can teach us – that, in contrast to the outcomes described in Chapter 3, the pace of evolution slows in environments with little light.

Steve Keable set about describing the isopods caught in shallow waters. There were clearly many new species – the contents of the shallow-water traps could be easily sorted into groups based on appearance. Museum volunteers without previous experience of either isopods or taxonomy could carry out this task. There were many obvious characteristics separating species A from species B. Some species had legs covered in spines, some without spines. Some had long antennae, others short antennae. And so on. The identification, and consequently the taxonomy, was straightforward for the shallow water isopods, but enhanced by Steve's refined taxonomic methods characteristic of the Lowry group.

To summarise, in shallow water evolution had resulted in many species of isopods, partly in response to the increase in niches created by light. And each species was considerably different – many genetic mutations had taken place over a limited time period, so evolution had been rapid where light levels were high. But how can I talk about time here, when all we have to examine are the species alive today? Surprisingly I can offer some justification. My evidence derives not from the fossil record of isopods – unfortunately that is inadequate. Instead clues can be drawn from the history of the Earth – plate tectonics, as described in Chapter 2.

The Australian plate is part of the Earth's crust. It consists of terrestrial land, and the submerged continental shelf and continental slope. The continental shelf inclines gently from the sea shore to a depth of about 200 metres. Then the continental slope commences as the sea floor plunges rapidly towards the Abyssal Plain, another gently sloping part of the sea floor, beginning at about 5,000 metres in depth. The base of the continental slope marks the edge of the Australian plate. So animals living on the sea floor down to depths of at least 1,000 metres are obviously well separated geographically where they occur on different plates. A species could conceivably occupy a large part of one plate, within a range of depths, by circumventing the land. But animals cannot migrate to other plates. They are divided by deep ocean, or forbidden territory. However, as described in Chapter 2, the different plates of today were once joined, but became separated throughout geological time. The consequence of this for animals is that species separated geographically today evolved from ancestors once living together on the same plate. It's also interesting to point out in this chapter that the Australian, Indian and Mexican plates (or continental slopes) were completely separated 160 million years ago.

Scavenging isopods were once caught during some early random trapping in Indian and Mexican waters. Steve Keable compared his shallow-water Australian isopods with these species. Just as there were considerable differences between each species within Australia, the scavenging isopods from India and Mexico were very different again. They were all related, in that they belonged to the same small branch of the evolutionary tree, but they had diverged considerably, to adapt to different niches in different light environments. So what can we learn from this?

The global picture informs of considerable evolution over 160 million years in an environment with substantial sunlight. One hundred and sixty million years ago, a population of ancestral isopods was divided geographically, travelling in different directions on board the continental shelves of three different plates. The ancestral species continued to evolve, but in three different environments. The result is that evolution yielded copious species in each case, but was different each time. Two environments are never the same, and evolution is reflected in this. But remember that here we are dealing with environments where light is present. In contrast, Steve's clearly defined mission was not to be echoed in Jim Lowry's task.

Jim was left with *Bathynomus* to tackle. At first, this appeared to be a prize project – *Bathynomus* was a magnificent animal. Then problems started to arise. The *Bathynomus* collected from each depth range

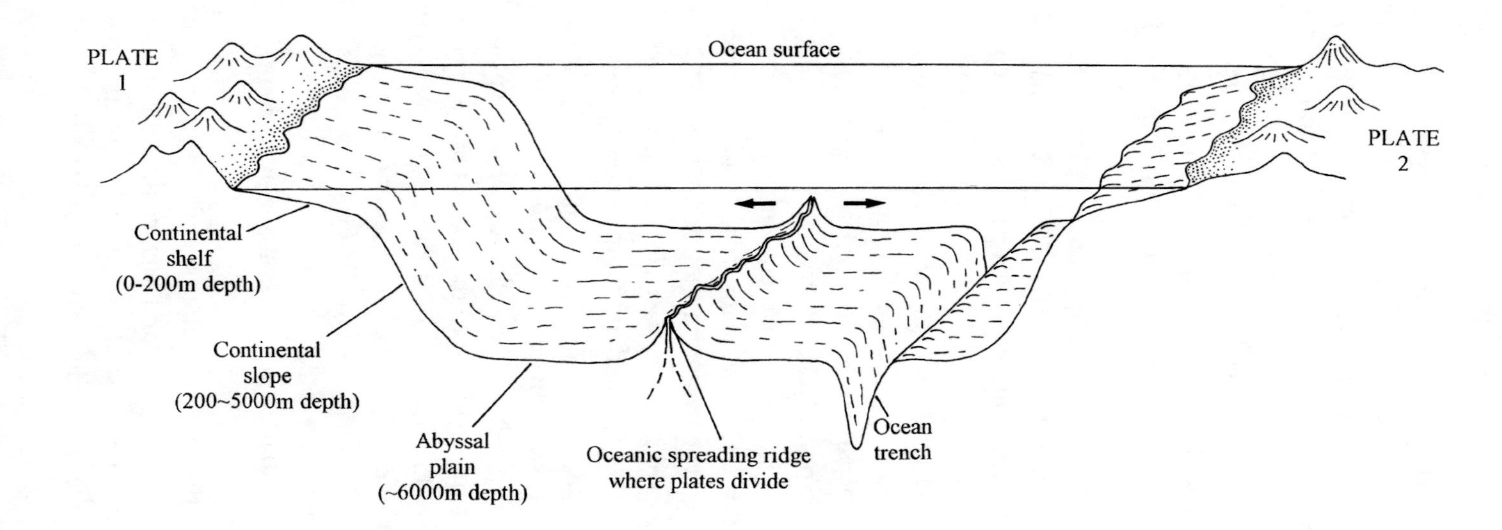

Figure 4.3 Simplified schematic section through two of Earth's plates, showing the submarine landscape between, including their line of separation.

on the Australian plate, beginning at 200 metres, all appeared similar. Those marked differences belonging to the shallow water isopods, obvious even to the inexperienced eye, were simply not there. There were slight differences – some individuals had four spines on a leg where others had five – but were these enough to designate more than one species of *Bathynomus* in the Australian fauna, and indeed, were there *any* new species here at all? These questions were at the foundation of Jim's taxonomic task, and the answers lay with the *Bathynomus* of India and Mexico.

A 'species' can be considered a group of similar individuals that reproduce in their natural environment. The word natural is important – related but different species can sometimes reproduce in an artificial environment, but would not do so under natural conditions. Of course, we could not observe the mating behaviour of *Bathynomus* at 1,000 metres. But when enough physical characteristics are recognised to reinforce a particular relationship, this can provide evidence towards classification. The characteristics of the other legs of the Australian *Bathynomus* were consistent with those of the first leg considered. Maybe this was grounds for designating two separate species from Australia. Maybe evolution had been slow in the case of the Australian *Bathynomus* over the last 160 million years – genetic variation was obviously very limited. It was time now to turn our attention to India and Mexico.

From fossils, we know that *Bathynomus* also existed earlier than 160 million years ago. It had travelled on separate plates, diverging from an original supercontinent to the regions that are now Australia, India and Mexico. In other words, it had not evolved from shallow-water isopods independently in all three locations during the past 160 million years. An examination of the *Bathynomus* caught by fishermen from India and Mexico would inform us what had happened to the ancestor over that 160 million years of living in different, isolated environments.

The pattern of the shallow-water isopods was not replicated in the deep. The *Bathynomus* of India and Mexico did not differ greatly from those of Australia – they were almost identical. Almost, but not quite. There was a size difference – although all were huge by isopod standards, exclusive size ranges were identified. These echoed the slight differences in shape such as spine numeration. But quite categorically, *Bathynomus* showed little variation in shape between species. It lived in deep water with little sunlight . . . and it had hardly evolved at all over 160 million years. *Voilà!* The evidence we have been looking for, and the point of the whole SEAS story.

This story paints a picture of what happens in environments with little light compared to those environments with considerable light. But the otherwise *x*, *y*, or two-dimensional spatial picture, has a third axis – *z*. The *z* axis represents time. And the complete picture is of restricted evolution where light is reduced. Genetic mutations have been diminutive as a result of modest selective pressures – pressures where light is not dominant.

Just to confirm that light is a major limiting factor here, we can compare the fauna living *within* the sediment of the sea floor of shallow and deep regions. Below the surface of sediment there is no sunlight. So a very different ecosystem exists there, a system not adapted to light. We have always known that the fauna of shallow water sediment is reasonably diverse, where most species derive from ancestors in the exposed waters above, but ecologists had predicted the opposite for deep-sea sediment. Then, in the 1960s, scientists from the Woods Hole Oceanographic Institution in Massachusetts collected deep-sea sediment samples using newly developed equipment. This technology was capable of collecting more specimens from a given area than ever before. And what it collected was beyond all expectations.

Although there were fewer individuals in the deep-sea sediment compared with its shallow-water counterpart, the number of species was similar. The diversity of life in deep-sea sediments was equal to that in the shallows. So a diversity of animals *can* potentially survive in the deep sea, and evolution *can* be as prolific as in the shallows – temperature and pressure, for instance, are not necessarily limiting to speciation. But where animals are adapted to sunlight, and the light levels fall, then the evolutionary brakes are applied and diversification slows down. The potential niches available diminish drastically. Armed with this clue towards solving the Cambrian enigma, we can leave the deep sea.

Now that we have adapted our vision and thinking to the dark, we are ready to examine an environment that is in total darkness. Rather than choosing the Abyssal Plain, I will select an environment that is slightly more accessible, and consequently one whose inhabitants are better known. Can we strengthen the message taken from the continental shelf and slope as light is removed from the equation completely? The answer to that question follows.

Caves

In his book *Colours of Animals*, Sir Edward Poulton devoted a certain amount of space to cave animals. He stated unequivocally that animals living in darkness were pale because pigment would not be visible in these situations and so would no longer be of any use to the animals. Poulton strongly favoured what became known as the Darwinian view of colour – that 'wherever colour is seen, it is due to the favouring influence of natural or sexual selection'. That Darwin carefully chose the words '*Whenever colour has been modified for some special purpose*' seemed to have been overlooked. So it is not surprising that Poulton extended his argument into environments without light. He suggested that in caves 'it [pigment] is, therefore, no longer maintained by natural selection, and *therefore* it disappears'. The second *therefore* became the subject of great dispute.

Another biologist of the time, J. T. Cunningham, believed that pigment was produced directly by the action of light on the skin. So he thought cave-dwelling animals were pale coloured because there was no light to stimulate the development of pigment. According to Cunningham, light and pigment were directly related. According to others, light is not the cause of pigmentation; it only puts in motion the machinery produced in the animal by natural selection.

Today, armed with genetic theory, we understand that Cunningham was wrong. But does the pigment machinery, or rather the process of genetic mutation and new gene deployment, stop working when light is removed? Are the cogs in this colour machine literally solar powered, in that they cease to turn without sunlight? Maybe the pigment machine has a reverse gear, one that is engaged in the absence of light. To uncover the complete story, we too should look into the caves.

It has been worthwhile examining environments with gradually decreasing levels of light, from the dim night-time of land to the almost complete darkness of the deep sea, if only to compare the communities which inhabit them with those of caves. In caves a similar transition in light levels exists. But here the transition occurs much more rapidly. Light fades away in caves over metres, rather than hundreds of metres as in the sea. And at the end of the journey into many caves, we reach a true, undeviating condition of total darkness on Earth.

I first became interested in caves when Mike Gray, an arachnologist at the Australian Museum, allowed me to examine his latest find. Mike had recently been underground in the Nularbor Plain of South Australia. Soon after entering the cave, he found himself in total darkness. And the fauna, visible under torchlight, rapidly became less diverse as his journey continued. But Mike found what he was looking for – a spider. More than that, he found a new species of spider. It's not unusual to find a new species of spider in Australia – Mike's previous discovery was made in his own garage. But this cave dweller appeared different to his garage specimen, or indeed any other spider living outside caves. It was related to the infamous 'Sydney funnel web' which meant it was supposed to have either six or eight 'eyes'. But with the aid of a microscope it became clear this cave species, just 15 millimetres long, had *no* eyes.

The deep-sea animals I had examined were adapted to even the most minuscule quantities of light present in their environment – they had big eyes. The cave spider was denied *any* light and had given up the evolutionary struggle to see. Its lack of 'eyes', nonetheless, *was* an adaptation to light. But did this 'eye' loss take place quickly through time? And how powerful was the selection pressure to lose 'eyes' a negative evolutionary response to light? It is difficult to answer these questions taking the cave spider as a model – we know too little about its relatives. But cave fish have been studied more intensively, and we have enough pieces of their puzzle to trace their journey through time, from the open ocean and into caves.

Sometimes bioluminescence exists in caves as it does in the deep sea – cave animals can produce their own light, like living torches. This, again, makes matters complicated – to begin with, we no longer know the exact light conditions. Bioluminescence may create an effectively continuous light field, or it may be intermittent. The light field may be relatively bright, dim or varied to any extent. Although bioluminescence probably causes a fairly faint light field within the big picture, it is best at this stage to consider just those caves where bioluminescence is absent, where the condition of total darkness is satisfied. Such a situation exists within the marine caves of Mexico.

Most inhabitants of marine caves today originate directly from their ancestors in the open sea. Either these ancestors now no longer exist, or they have moved into some other extreme environment. For instance, one group of small crustaceans, called remipedes, is virtually confined to cave habitats today, even though their evolutionary origin was in the open sea. They are known as a relict fauna – species derived from groups that were formerly widespread and diverse but now survive exclusively in a cave, possibly, according to Bermudan cave biologist Thomas Iliffe, because of reduced competition or predation. Iliffe found that some remipedes in eastern Atlantic caves look very similar to those in caves of the western Atlantic, and this similarity was not the result of convergence – the evolution of similar bodies to adapt to similar environments. Instead the similarity signalled almost zero evolutionary activity. The caves have been separated geographically for over 100 million years and, as for *Bathynomus*, very little evolution had taken place in the dark environments. Even closer to the *Bathynomus* story, isopod crustaceans found living in caves, isolated for over 100 million years, also bore a remarkably close resemblance to each other. In fact this story is echoed in many types of animals. And in most cases the explanation given for their current cave living is the same – that their ancestors once inhabited shallow, open seas but were driven out by competitors and predators among the new faunas that appeared throughout geological time. But the Mexican cave fish can provide more information. They have a very close relative living outside their caves today.

In the previous chapter we saw how the angelfish employs its silvery surface for reflecting light at its opponents, in the style of *Star Wars*.

But there is another, more widespread function for the silver colour of fishes – to make them disappear.

In near-surface waters, such as the angelfish's Amazonian habitat, sunlight exists in the form of a beam like a spotlight, as it does on entry through the Earth's atmosphere. But below these waters the beam formation is broken, and sunlight is scattered in every direction. So here objects are illuminated equally from all directions, and no shadows are cast. A mirror in these waters vanishes from sight because in the mirror one sees only a weak reflection of the environment. The mirror becomes an optical illusion – in the direction of the mirror there appears to be only the background environment, with nothing in the way. In the ocean a silver fish is effectively a mirror. A predator looking directly at a silver-sided, or mirrored fish from below sees only a reflection of the surface. So in the direction of the fish there is . . . no fish! But how can a fish's skin act as a mirror? After all, it contains no metal. There is another way of strongly reflecting *all* the colours in sunlight into a beam so that it appears as a very bright white, which we know as silver. We turn to structural colours.

In Chapter 3 we learnt that a thin film causes colour – structural colour. Also, a stack of thin films was found to provide a relatively brighter colour, by reflecting a greater proportion of sunlight. But the reflector caused strong coloured effects rather than white because the thin films were all of the same thickness, and this thickness determined the wavelength, or colour reflected.

Now imagine a stack of thin films of different thicknesses. Imagine some that reflect blue light above others, that reflect green above yet others, that in turn reflect red. As sunlight strikes this structure, its blue rays would be reflected from the top layers, leaving the green and red rays to continue along their original path. As these rays meet with the middle layers, the green rays are reflected, leaving only red rays to continue along their path. And finally the red rays meet with the lower layers and they too are reflected. So the combined effect of all the layers is the reflection of blue, green and red rays in the same direction. And blue, green and red combine to form white, or silver (silver is a strongly directional form of white). With more layers of different thicknesses, more colours in the spectrum can be reflected. And this is how the fish skin appears silver – it contains a stack of layers of varying thickness.

In the Sierra Madre Oriental mountain range of eastern Mexico lives *Astyanax mexicanus* – a fish some 5 centimetres long commonly kept as pets in domestic aquaria. It is related to South American piranhas. In open waters this fish has average sized eyes, for a fish, and a silver body to provide effective camouflage. I will refer to this form of the species as the eyed cave fish. Its eyes and silver colouration are obvious adaptations to light. The same species of fish also inhabits the extensive cave systems of Mexico, but as a different form . . . or rather forms.

As the eyed cave fish moved deeper into the cave system through geological time, the selective pressures to be adapted to light vanished. And as they did so, the structures and chemicals – the hardware and software – of the animal responded. The eye began to degenerate. The longer the cave fish spent in darkness, equating to the further into the darkness the fish ventured, the more the eye degenerated. The evolutionary machinery had not stood still, but it had engaged reverse gear. 'Regressive evolution' was the trend as far as light was concerned. The adaptation to light outside the cave had resulted in some expensive hardware and software. Within the cave, the energy of the fish could be put to better use. The visual machinery, which had become obsolete, had to be dismantled. And it was not only the eye that regressed – the silver colouration was affected, too.

At Oxford University, Victoria Welch studied cave fish from within the vast Mexican cave system. She noticed that the fish were becoming less silvery as their habitat moved deeper into the caves. And as the silver disappeared, so their skin became a translucent white colour, with the red of their blood vessels creating an overall pink effect. But the transition from silver to pink was a gradual one, with intermediate forms appearing as an unbalanced collage of both states. This, however, was not the only pattern to emerge.

The eye was absent from all forms of cave fish living in the dark caves. It has undergone regressive evolution rapidly – the eye is a *very* expensive piece of equipment, and one that must be relinquished the moment it becomes obsolete. But the silver colouration turned out to be a little cheaper in terms of energy investment. In fact the silver colouration may also have been influenced by 'genetic drift' – mutations that just happen under neutral selective pressures.

Cave fish populations found deeper within the caves had been living in the dark for longer, in geological time, than those populations living nearer the cave entrance, albeit still in complete darkness. And since it took longer for the silver colouration to regress compared with the eye, the cave fish near the entrance of the caves were more silvery than those in the deepest parts of the caves. In fact the fish furthest inside the caves were completely pink.

Victoria questioned what was happening in the skin of these cave fish. How was the silver reflector being effected? She took samples of skin from fish at different depths within the cave . . . and found the cause of the silver decline. Evolution was observed mid-action.

In an electron microscope, the individual thin films, or layers of the silver reflector, can be observed. The eyed cave fish possessed very ordered stacks of layers, which increased gradually in thickness from the blue to the red reflectors. In those fish living near the entrance of the cave, but in the dark, signs of disorder began to show. The layers were beginning to separate, split apart and even become fewer in number. As the fish found from deeper within the cave were examined, these signs of disorder became more pronounced, and the total number of layers gradually reduced. The layers also began to buckle and became randomly distributed within the skin, and the skin became less silver. Eventually, in the fish from the very depths of the cave, the layers had vanished completely from the skin. There was no longer any reflector.

This was a nice find – the different stages of regressive evolution could be observed happening through time. If a silver reflector became obsolete within a sunlit environment, this event would be rapid and impossible to track. The cave finding may also indicate how silver reflectors *evolve* in the first place, possibly by reversing the procedure. But the real moral of this story, for the purposes of this book, is once again that evolution may take place slowly in an environment without light. Indeed, the cave fish had not evolved sufficiently to form a new species during its long history of entering very different environments – all without light.

The lack of light in caves resulted in reduced environmental partitioning into microenvironments – quite the opposite to the case of the West Indian *Anolis* lizards. Consequently, the island-type evolution that is encouraged by microenvironments was absent. The outcome was a lesser variety of species although still a considerable number of individuals in caves. The question I will pose later in this book is: 'Was the Precambrian environment similar to the modern cave environment?' We can start to think about this question here, making the clues for solving the Cambrian enigma to be found in the following four chapters appear all the more relevant.

Other experiments have been conducted to show that animals inhabiting dark caves are completely unaffected when light is shone into their surroundings. So they really have become visually neutral. In fact a number of cave animals have been found in illuminated habitats where *no competitors from the surface had access*. They are *never* found in similar habitats that do contain competitors or predators adapted to light, because if they stray into these environments they do not survive for long.

In Chapter 3 I mentioned that many deep-sea animals are red coloured, and that this was an adaptation to light. There is one shrimp that exists either within or at the entrance of deep marine caves. It changes colour from a pigmentless white to red, as it moves from within the cave to the caves' entrance, where light exists. The adaptation to light is significant *everywhere*. Also in Chapter 3, we compared (as we have to some extent in this chapter) the senses of smell and taste, hearing and touch with vision. It was concluded that vision is different because its stimulus, light, was always present in the environment. Every animal in that environment is affected by light. In caves these other senses are extremely well developed, yet evolution labours in first gear. Animals are to some extent in control of how much sound and scent is injected into the environment, but in a sunlit environment the light levels are pre-set.

Darkness is the most obvious characteristic in the caves considered in this chapter. It acts directly on animals by placing blind species at no disadvantage to others. But it also has an indirect action – it excludes photosynthetic organisms, thereby reducing the amount of locally produced food to zero. This nutritional poverty will affect the cave food web, but it should not affect biodiversity, or the evolution of species, as much as the number of individuals, or density of life. And it is the evolution of species that is most relevant to this book. Indeed, most cave predators have adapted to go without a meal for weeks, even months.

Despite the fact that cave environments are remarkably stable, lacking extremes of anything, and that senses other than vision are remarkably well developed in the dark, diversity in caves is low. Evolution is slow. And this can be attributed to the lack of light to fuel both photosynthetic organisms and vision. Often in this book I have referred to 'light' *and* 'vision'. Soon I will discriminate judiciously between the two. Light has existed on Earth from its very beginnings. Vision is an adaptation to light. It has not always existed. This is worth thinking about.

Vision will be dealt with exclusively in Chapter 7, but first we will move out of reverse and examine what happens as the *forward* visual gears are engaged in the evolutionary machine, in the case of the luminous seed-shrimps.

5

Light, Time and Evolution

Life abounds with little round things

LEWIS THOMAS

Ostracod crustaceans, or seed-shrimps, have travelled through time well. They are abundant today and were equally common throughout the past, right back into the Cambrian period. They are found in all types of water worldwide, and their poor public exposure is not reflected by the extent of scientific attention they have received. Around 40,000 species of seed-shrimp have been described – rather significant, considering we know of only about 8,700 species of birds and 4,100 species of mammals (although this is more in line with some of the other highly diverse invertebrate groups). But when the name 'seed-shrimp' is spoken, the conversation generally refers to just one group of seed-shrimps – Podocopa, species with generally thick, robust shells. I will refer to Podocopa as the 'heavyweight' group. The bias towards heavyweights has been generated by palaeontologists – heavyweights can be used to indicate the presence or absence of oil reserves – but in this chapter the other side of the story will be heard. It is another group of seed-shrimps that will contribute to the Cambrian enigma. They will introduce the subject of colour to that of animal evolution – a relationship which will be seen to flourish as this book progresses.

Seed-shrimps, like scallops, possess a two-part shell that can enclose the entire body, although typically the shells of heavyweight seed-shrimps are only a millimetre long. Heavyweights owe their popularity to their shells – the shell chemicals are fossilisation friendly. Consequently they have left an extensive fossil record. Palaeontologists have kept a good eye on the movements and activities of the heavyweights throughout geological time, spurred on by a dangling carrot. There *is* 'gold' at the end of this palaeontological rainbow. Heavyweight seed-shrimps are well-known indicators of oil reserves, and until the recent introduction of more sophisticated oil detection methods, the laboratories of oil companies bulged with heavyweight seed-shrimp specialists. There exists, however, another group of seed-shrimps – Myodocopa, species with generally less robust shells. I will refer to the Myodocopa as the 'lightweight' group. Lightweight seed-shrimps have a different form of the chemical that constitutes their shells, and this form does not usually give rise to fossils. So for some time we were unsure about the historical whereabouts of the lightweights.

In the early 1980s, David Siveter, a palaeontologist from Leicester University in England and part of Chapter 2's 3D fossil reconstruction team, fractured a rock he had collected from Scotland. The rock was around 350 million years old. Inside it were fossils, oval in shape with a tiny notch at one end, and totalling around 5 to 10 millimetres in length. Could these be seed-shrimps? The shape suggested yes, possibly, but the size no. Not all groups of living seed-shrimps, however, were well understood. And before comparing the Scottish fossils with living species, we need to know exactly what is out there in the water today.

The SEAS expedition did achieve its target – representatives of the scavenging amphipods and isopods were collected successfully. The 'pods had been gathered. But, surprisingly, they were not the most abundant groups of scavenging crustaceans. Another group of crustaceans emerged as the scavenger supremo of eastern Australia – the 'cods. Ostracods – seed-shrimps. This situation was highly irregular – seed-shrimps were not thought to hold a position of any note in the hierarchy of the world's scavengers.

The seed-shrimps that happened to like frozen pilchards and wandered into the traps were the lightweights, the group that had left little

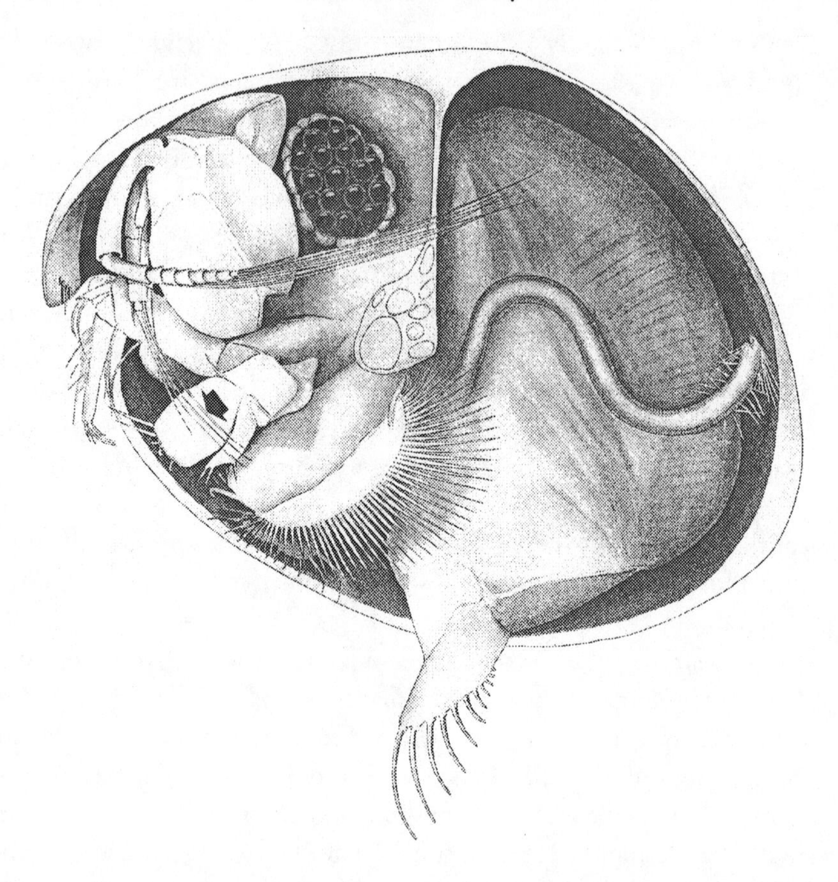

Figure 5.1 A notched lightweight seed-shrimp with one half of its shell removed to reveal its body and limbs inside (from Cannon, 1933, *Discovery Reports*). The arrow points to the halophores of the left first antenna.

behind in the fossil record. And in particular it was just one family of lightweights – Cypridinidae, seed-shrimps that generally have a small but well-defined notch at the front of their shells. I will refer to the Cypridinidae as the 'notched' group. Notched seed-shrimps are usually the size and shape of tomato seeds, and typically spend much of their time buried in the sand on the sea floor. The tomato seed lookalikes occurred in the traps set in shallower waters. They were common also in traps set in deeper waters, but at depths of 200 and 300 metres they were occasionally accompanied by an oddball among notched seed-shrimps – the 'baked bean'.

The very first trap set in a depth of 200 metres off the coast of Sydney was hauled up and opened on board the fishing vessel hired for the job. The sight was as amusing as it was unnatural – the trap was full of what appeared to be baked beans. 'Baked bean' is the official nickname given by local fishermen to 'giant' orange/red seed-shrimps called *Azygo-cypridina*. Sometimes they are brought up in the catches of fishermen, who have no idea what they are dealing with, merely that they 'don't make good eating'. Baked beans appear to be confined to the edges of the continental shelf. Like real baked beans, these seed-shrimps are about a centimetre long, oval and slightly flattened from side to side. They are orange/red because at a depth of 200 metres and beyond sun-light is almost exclusively blue. You can be sure, at least, that at such depths orange and red will be absent from the sun's spectrum. And with nothing left to light up the baked beans, they appear to be invisi-ble. In a totally dark room, for instance, an orange cannot be found with a blue torch. But there is a difference in the appearance of the deep-water seed-shrimps and baked beans – the seed-shrimps possess their characteristic notch at one end. And so did the Scottish fossils.

Living fossils

Every once in a while a 'living fossil' is discovered somewhere on Earth. Living fossils are species alive today that closely resemble forms other-wise found only as fossils, species that lived during ancient times. The nautilus could be considered a living fossil because it shares its looks, behaviour and, more importantly, its place in the evolutionary tree with the extinct nautiloids and ammonoids. But the nautilus lives on.

One of the most recent living fossils to emerge is the Wollemi pine. Fossils of this type of conifer were once known only from rocks con-taining dinosaur bones, and it was thought to have been extinct for millions of years. It was an important subject in palaeontology. Then all the hard work put into extracting and extrapolating the fine details of its anatomy to bring it to virtual life was undone by a single event – the dis-covery of a living specimen. Virtual life was suddenly replaced by real life – an occupational hazard for palaeo-artists (although a rare one).

In a remote part of New South Wales, Australia, forty adult Wollemi pines were found very much alive in a deep, sheltered gorge. The gorge supported a warm, temperate rainforest. It may seem amazing that the massive pines had not been discovered before, but much of inland Australia is actually unknown to science. There is considerable biology to be done in Australia. Spiders, for instance, are among the more populous animals on Earth, yet around two-thirds of Australia's spider species probably remain undiscovered and unnamed. So it is not so surprising that if a new species of tree is discovered today, it will turn up in Australia. Of course, as the world's rarest tree, the Wollemi pine must be closely monitored and protected, so much so that its precise location has been kept a secret. Even cultured specimens growing in Australian botanical gardens are kept under lock and key, and beyond the reach of the horticultural black market.

There is an obvious link between the Wollemi pine and the SEAS project. In the deeper localities, hagfish were caught in the large scavenger traps. Protected within a scabbard of slime, hagfish appear like eels. They have primitive mouthparts, and indeed are today's representatives of a primitive form of fish. Their mouthparts are quite an issue because hagfish have a strong fossil record, dating back some 500 million years, but the fossils show no sign of jaws. And the living hagfish confirm this – they really are jawless. The jaw is a feature of more derived forms of fishes – sharks and bony fish. But hagfish are scavengers, and can really get by without a jaw in certain environments, those in which they are preserved today. The relevance of the Wollemi pine and hagfish to this chapter is that the baked bean is a living fossil too. Although less apparent to begin with, morphometric analyses were employed to expose this truth.

Baked beans showed similarities in form to the 350-million-year-old, oval fossils discovered by David Siveter, who by this time had classified them as lightweight seed-shrimps. Morphometrics can give mathematical values to shapes. A morphometric value, in the form of relative coordinates on a grid, was given to the Siveter fossils, and it seemed appropriate to put the baked beans to a similar test. The match was perfect. David Siveter was right – he *had* found lightweight seed-shrimps that lived 350 million years ago. To be more specific, these fossils belonged to the notched group of lightweight seed-shrimps. Before long, David Siveter and his team discovered further fossil lightweight seed-shrimps, now that they knew what to look for.

Different forms of lightweight seed-shrimps were uncovered from older rocks, but here the baked bean forms were absent. The lightweight group as a whole could be dated back 500 million years to just after the Cambrian. But it seemed that the baked bean form, and the notched seed-shrimps in general, evolved about 350 million years ago. Now, for the first time, we were beginning to trace the geological history of the much forgotten lightweight seed-shrimps. But it was useful, also, to have a date on the evolution of baked beans for another reason.

Diffraction gratings – a subject of physics

The study of structural colours in animals has a long and distinguished history. Robert Hooke possibly pioneered the subject in the seventeenth century with his interpretation of the metallic appearance of silverfish insects, just pipping Newton to the post. And ever since, this subject has been famously represented, up until the work of Sir Andrew Huxley, Sir Eric Denton, Michael Land and Peter Herring in the latter half of the twentieth century. Consequently, animal forms of multilayer reflectors and structures that cause the scattering of sunlight, with all their variations, had been well documented and interpreted by biologists. But these were also subjects of optical physics. Physicists have been experimenting with optical materials for centuries, and had converged on the same structures that occurred in nature. Yet the two fields of biology and optical physics never seriously crossed paths.

Despite numerous studies on animals known to show metallic-like reflections, such as many beetles, butterflies, fishes and hummingbirds, there remained physical, optical structures that were known to physicists but not to biologists. Prisms, for instance, had never been found as light reflectors in animals. Perhaps their precise shapes or copious volumes made prisms an evolutionary impracticality. 'Prisms' can be found occurring naturally, nonetheless, in raindrops that refract and reflect sunlight to create a rainbow.

In 1818, another type of physical structure with reflective properties was invented in a physics laboratory – the diffraction grating. Fine copper wire was wound tightly around a screw, and the acutely grooved surface created by the wire caused sunlight to be split into its component colours: a spectrum was reflected. A different colour could be seen from different directions. Diffraction gratings could be considered as tiny corrugated sheets, where the spacing of the grooves are fairly constant and approximate to the wavelength of light. At their most efficient they are microscopic. Diffraction gratings became major players in the scientific and commercial worlds of optics, and have become refined and varied to produce an array of optical effects. They are responsible for the metallic-like, coloured holograms found on credit cards or foil-type wrapping paper, and now they are also being used on stamps and banknotes since they are difficult to forge. But they were unknown in nature and the subject of animal structural colours until 1993.

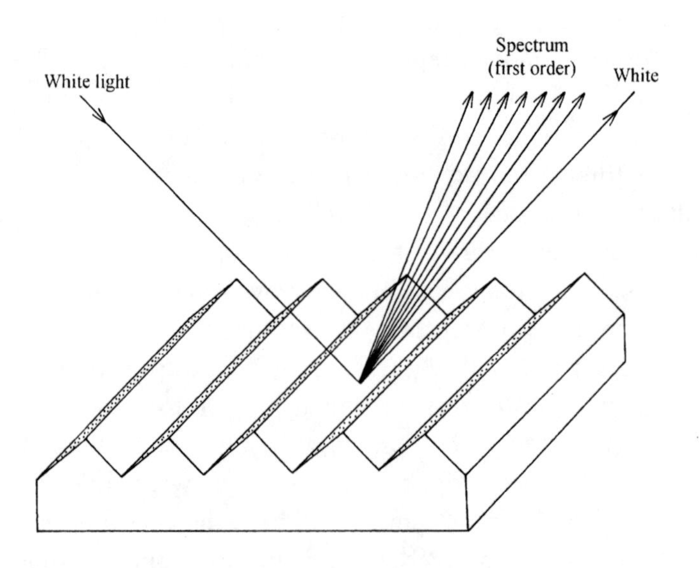

Figure 5.2 A diffraction grating splitting white light into a spectrum.

A sudden flash of green light

My role in the SEAS project was to describe the new species of seed-shrimps collected. Sixty unknown species of notched seed-shrimps emerged from an area where only a couple of species were thought to exist in total. But it was not simply their diversity that made the notched seed-shrimps such important scavengers; it was their abundance. A single trap, basically a foot-long section of drainpipe baited with a couple of dead pilchards, would attract up to 150,000 individuals. Considering the short distances seed-shrimps are prepared to travel for food, the SEAS findings indicated that notched seed-shrimps were probably the commonest multicelled animals on the Australian continental shelf. Yet until then they were virtually unknown. This typifies how little we know of the smaller, but probably more common, life forms on Earth. But thanks to the SEAS project, the secret of the notched seed-shrimps had been revealed. Well, at least the secret of their Australian affluence. A further secret lay waiting to be discovered, one that could only be revealed using microscopes.

To examine the body parts of preserved seed-shrimps, their shells must be removed. This operation involves manipulating a specimen under a microscope and attempting to sever the muscles that hold the shell closed. The tiny size of most seed-shrimps makes this job difficult, and often several attempts are needed. The seed-shrimps tend to roll around and fall in exactly the positions that are not required of them. One exceptionally long day in the Australian Museum I had been battling with seed-shrimps for this very reason. It was time to go home but I was delayed. Then something happened that would change the course of my research – *I saw a flash of light.*

As one preserved seed-shrimp rolled over in the glass dish under my microscope, it caught the microscope's light and sent an extremely brief blaze of green light towards my eyes. Unsure of what it was, or indeed if I was seeing things, I rolled the specimen over again, in an attempt to repeat the performance. Once again it shimmered with green light. Holding it in the appropriate position, the green reflection shone continuously. The shell of the animal appeared rather dull and its background was decidedly black, but the green light was blazing like a neon sign in the night. I asked my nearest companions, the amphipod specialists Jim Lowry and Helen Stoddart, to double-check that this was really happening. It was, but it shouldn't have been. There was a big literature on seed-shrimps, and green flashes were not part of it.

The green part of the seed-shrimp belonged to its first pair of antennae. These antennae are equipped with long hairs, and each long hair is the bearer of smaller hairs, called halophores. Halophores are flexible because they are made of minute rings, stacked side by side. They are held together by a thin, elastic outer skin. But like the fine wire wrapped around a screw, they cause tiny ridges and grooves to appear on the outside of the halophores. The light microscope indicated that the green flash came precisely from the halophores. The electron microscope revealed the spacing of the very regular grooves – it approximated the wavelength of light. The surface of a halophore was a diffraction grating. Again, it shouldn't have been. There was also a considerable literature on structural colours in animals, and diffraction gratings, like seed-shrimps themselves, were absent from it.

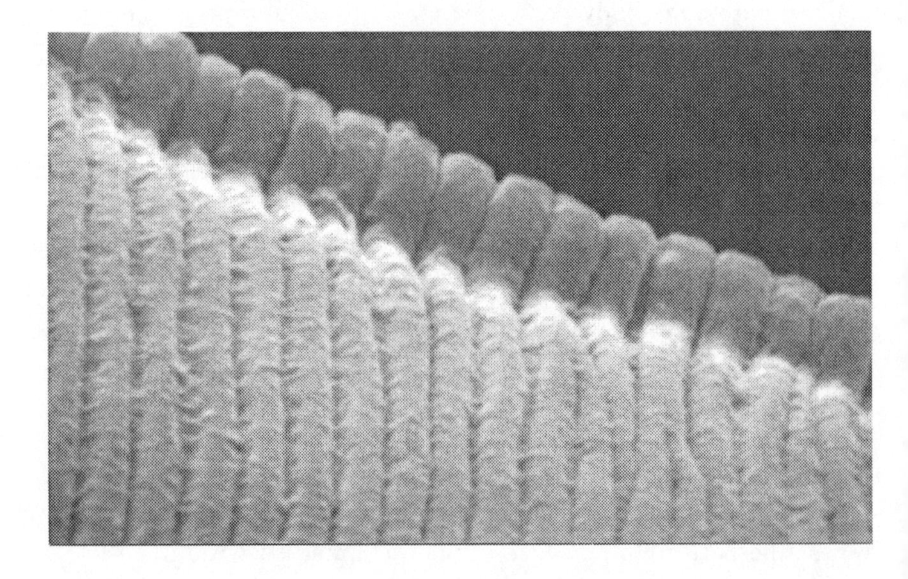

Figure 5.3 Scanning electron micrograph of a diffraction grating of the 'baked bean' (*Azygocypridina lowryi*). Spacing between grooves is 0.6 microns. (Plate 15 in the colour section shows the iridescent effect of this structure.)

Next, the halophores from a range of notched seed-shrimp species were examined. All possessed the iridescent character but to varying degrees. Some species reflected a spectrum of colours, others just green, just blue, or just blue and green. The electron microscope provided the source of these variations – the diffraction gratings were different. This was becoming interesting, but before investing further time and money in notched seed-shrimp iridescence, a great barrier had to be crossed. The question hanging over this work was, 'Does iridescence play a role in the lives of seed-shrimps?' This question was fundamental. If the answer was 'no', it was time to forget that the original green flash had ever happened. A colour that has no function must be purely incidental (I say incidental rather than accidental because everything that has evolved, even those things with a function, are accidental). And an incidental colour has no place in the literature of either seed-shrimps or animal structural colours. But if the answer to this big question was 'yes', it would be time to call in the optical physicists. So how does one find the answer to such a question, especially since there is so little background information on notched seed-shrimp behaviour from which we can start? Well, sometimes one needs some luck.

The feeding mechanism of notched seed-shrimps was unknown, but the SEAS project had elevated feeding to the top of the 'things to study' list. To be entitled to wear the crown of scavenger, notched seed-shrimps must have an efficient feeding mechanism. So when an opportunity arose to film notched seed-shrimps in action, it was grasped with both hands.

In 1994 a film crew came to town to record the marine life of Sydney. On a wharf just within the harbour, they constructed an impressive aquarium through which fresh seawater flowed continuously to create a deceptively natural environment. The run-of-the-mill anemones, starfish and crabs were introduced and conducted their business as usual, which was monitored in detail via a camera so large it must have been good. Certainly the highly magnified pictures on the monitor were impressive, as was the control system – the camera could be steered on tracks in all directions. And somehow the film crew came to believe, or were tricked into believing, that seed-shrimps would make compelling viewing, and seed-shrimps were hired as extras on the final day of filming.

Wasting no time, a scavenger trap was rushed from the Australian Museum to the local beach – Watson's Bay, within Sydney Harbour. The beach was 100 metres or so long, and there was time to target only one spot – what was hoped would be a seed-shrimp hotspot. The rocks that bordered the beach were the initial choice, until a fish and chip shop was spotted at the end of a wharf. Their degradable waste often ended up in the water: what better place to find scavenging seed-shrimps than on a pile of discarded fish carcasses? Notched seed-shrimp heaven had been found – the recovered trap was full of them.

The seed-shrimps in the trap were transferred to a large bucket of seawater and chauffeur-driven to the film set. They began performing well. Some were swimming at full speed while others were stripping a pilchard to the bones. According to the script, it was the eaters that would star. And it was good to find that one part of the seed-shrimp's body had evolved into a relatively large, saw-like tool that could slice efficiently through fish skin. But the show was stolen by two individuals on the surface of the bucket of water – they appeared to be mating. This was certainly not in the script. Notched seed-shrimps, or any lightweight seed-shrimps for that matter, had never been accurately observed mating. All that was about to change.

The pair were transferred to the big stage during the final hour of filming . . . and they mated, shells juxtaposed, lower surface to lower surface. It was nice to discover this, but the real cause for celebration happened just seconds earlier when the male seed-shrimp performed a courtship ritual. He circled the female then . . . *he released a flash of blue light*! His iridescent halophores had been withheld within his shell. Then, when he was in full view of the female, his halophores emerged from his shell in all their iridescent glory. And, like a peahen with the tail of a peacock, the female seed-shrimp was suitably impressed – they mated. It was extremely fortunate that a single pair of notched seed-shrimps had chosen *that* particular moment to mate, and with only an hour's worth of film remaining. This was just lucky.

The discovery that iridescence was employed by notched seed-shrimps changed everything. Rather than ending up as a footnote in some obscure publication, notched seed-shrimp iridescence could now be taken seriously. It was time to alert the physicists. The species captured on film was *Skogsbergia*, named after an early seed-shrimp specialist. This notched seed-shrimp displays exceptionally spectacular iridescence, but in the males only. The females are quite dull in comparison. And this difference between sexes became clear in the electron microscope.

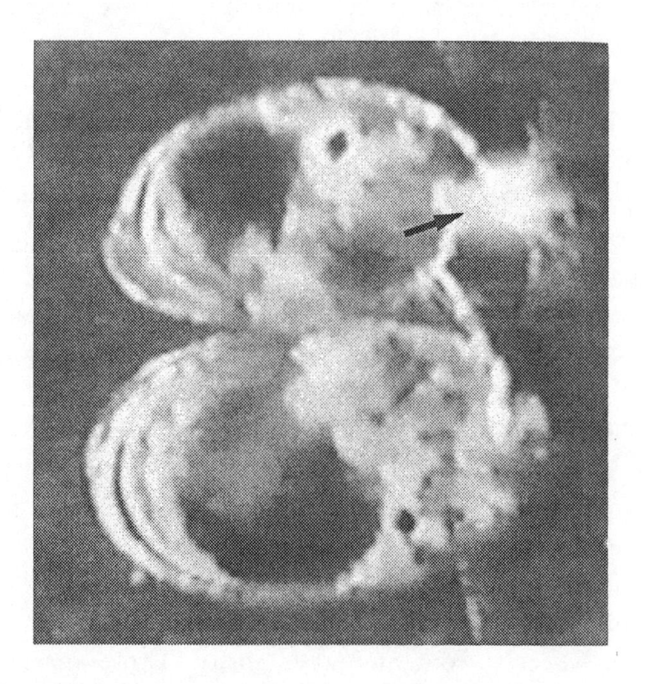

Figure 5.4 Frame from a video recording of a pair of the notched seed-shrimp *Skogsbergia* species mating. The iridescent flash of the male is arrowed.

The antennae of male and female *Skogsbergia* were coated in a thin layer of gold and then bombarded with electrons, rather than light. The images formed from reflected electrons revealed male antennae swamped in the iridescent halophores, and sparse halophores on female antennae. At higher magnifications, differences between individual halophores emerged. Male halophores have the profiles to cause an optimal reflection of blue light at both macro- and microlevels. In terms of optics, they form extremely efficient diffraction gratings. The diffraction gratings of female halophores, on the other hand, are decidedly crude. This conclusion was reached following collaboration with physicists. Optical physicists employed their rigorous electromagnetic scattering theory on the iridescence of *Skogsbergia*, followed by that of other notched seed-shrimps. A pattern gradually began to emerge.

Different species of notched seed-shrimps possessed different iridescent properties. Optical efficiency values were given to all of them and the possibility arose that they could be placed in a sequence, in order of iridescent effectiveness. Efficiency values were calculated using both the physics of the diffraction gratings and the design of the halophores on a larger scale. The values derived from many components – so many that a more sophisticated sequence could be constructed using cladistic methods.

Cladistics is a mathematical method for calculating relationships between species based on a character set – each species on Earth has an individual set of characteristics, both structural and genetic. The relationships are illustrated in the form of a family tree, and the family tree can be used to suggest an evolutionary tree. Cladistics is a common tool in the study of evolution, and in this case it did indeed generate a neat sequence of species based on their iridescence. And the increasing sophistication in structure of halophores was mirrored by their visual effects. The observed effect of iridescence also appeared to be transforming in spectral content and intensity. Those species at the beginning of the sequence reflected all colours equally, each colour projected in a different direction, while those at the end of the sequence were reflecting only blue light, and more intensely than ever before. Green and blue-green reflections lay somewhere in between. But what did this sequence infer? Did it in fact mean anything? Could there be implications for evolution here? The question of evolution was confronted first, and the work of specialists in bioluminescent seed-shrimps became appropriate to the case.

Bioluminescent seed-shrimps

Chapter 3 introduced the remarkable array of mechanisms that have evolved to provide colour in animals living under sunlight. Sunlight is reflected, transmitted and absorbed in all sorts of ways to produce different visual effects. The iridescence of the notched seed-shrimps considered so far is an example of one reflective effect. But what about those animals living without sunlight? This case was examined in the previous chapter, but only in part. Actually there remain animals living without sunlight, such as some deep-sea or nocturnal species, that are extremely visual. The animals referred to in Chapter 4, that were evolving modestly, did not employ light to any notable degree. So how do you operate with light in the absence of a light source? Quite simply, you make your own.

Seed-shrimps cannot generate electricity to power miniature light bulbs. Instead they adopt a more efficient method of yielding light – they bioluminesce. Two chemicals – a luciferin and a luciferase – react with the oxygen in water, and light is emitted as a byproduct. The light is referred to as bioluminescence. Only about 20 per cent of the energy fed into a light bulb fuels light; the rest is lost as heat. Bioluminescence is less wasteful – almost all of the energy investment becomes light, and so it is known as 'cold light'. Luminescence can be seen at fairgrounds at night. Plastic tubes containing luminescent chemicals, separated by a thin glass wall, are sold as necklaces for children. When the plastic tube is bent, the inner glass wall is broken. The chemicals mix and the necklace glows in the dark like a neon sign (although neon signs themselves employ a different mechanism). Similar plastic tubes are sold to Scuba divers and fishermen for conducting their business in the dark. It is easy to find a luminescing diving buddy in the water at night, and a fishing float that luminesces is just as conspicuous in the dark. Sometimes the plastic tubes are unnecessary for these marine activities; natural bioluminescence can suffice.

Waters abundant with bioluminescent dinoflagellates – single-celled organisms – can make swimming or diving at night an extraterrestrial experience. As the movement of arms and legs agitates the dinoflagellates, they react by mixing their luminescent chemicals. And the effect is so powerful that a sharp human silhouette is clearly visible as a blue or green glow in the dark. In fact the Australian Navy are clearly concerned because they closely monitor the geographical movements of their bioluminescent natives. No matter how cleverly you design your ship, if it sails into a crowd of bioluminescent dinoflagellates it lights a beacon to let everyone know where you are. And lightweight seed-shrimps have evolved bioluminescence with similar aspirations.

One group of lightweight seed-shrimps, the Halocyprida, produce bioluminescence from organs in their shells. I will refer to the Halocyprida as the 'eyeless' group, since all their representatives lack eyes. The two bioluminescent chemicals are pumped from eyeless seed-shrimps into the water, where they react to form a luminescent cloud. These eyeless species gather at the ocean surface at night and bioluminesce for all they're worth. The result is a mass 'light bomb'; a patch of bioluminescence so bright it can be detected by satellites in space. And the reason for this is so as to generate a burglar alarm. The seed-shrimps are eaten by small fish, and small fish are eaten by bigger fish. Any small fish entering the light zone becomes a most conspicuous silhouette, and the alarm bells sound for the bigger fish. Not surprisingly, the eyeless seed-shrimps remain undisturbed at night.

There is another group of lightweight seed-shrimps that bioluminesce, and these, as it happens, are the notched seed-shrimps. But only some of them are luminescent, perhaps half of all notched seed-shrimp species known. I found my first bioluminescent notched seed-shrimp on a beach in Australia. Although they were known to live there, all I could find was a luminescent crab, impressive as it was. A crab glowing intensely in the dark is an extraordinary sight, but it was not the crab itself that was glowing: it was its food. The crab, a transparent species, had eaten a notched seed-shrimp and the bioluminescent chemicals were mixing in its stomach. There have been similar reports of this happening in other parts of the world, so maybe bioluminescence is not such a problem for the crab.

Notched seed-shrimp bioluminescence evolved independently of eyeless seed-shrimp bioluminescence. The bioluminescence of notched seed-shrimps originates from organs in their lips. Katsumi Abe, a Japanese biologist from Shizuoka University, was a serendipitous explorer of notched seed-shrimp bioluminescence. He concluded that the chemicals responsible for this light evolved from digestive enzymes. This makes sense – the bioluminescent chemicals do share exit valves with digestive enzymes. And this could be a significant finding since the foundation of bioluminescent chemicals is a contentious issue of evolution. Sadly, Katsumi Abe died before the full extent of his research, or his thoughts, could be known. Fortunately his students, and his colleague Jean Vannier from the University of Claude Bernard in Lyons, France, are continuing along Katsumi's path.

The independent origin of notched seed-shrimp bioluminescence is echoed in its rudimentary function. Like the eyeless seed-shrimps that luminesce from their shells, the Japanese notched seed-shrimps also employ their light to counter predators. But the notched seed-shrimps endeavour to confuse rather than deter their predators. When a fish gets too close for comfort, and the seed-shrimp assumes it has been spotted, a blinding flash is created. The intense light briefly stuns the fish (just as we are often momentarily blinded by a glimpse of the sun), giving the seed-shrimp an opportunity to run for its life. And like a magician's assistant, when the smokescreen has disappeared, so has the seed-shrimp. That this feat is performed at all means it must be effective, because disadvantages are inherent in this strategy. A flash of light may curb the aggression of a prospective predator, but will also attract the attention of a more distant enemy – a flashing light is more conspicuous than a steady one. This antipredator response does appear to have been the original purpose for notched seed-shrimp bioluminescence, when it first evolved. But it has also provided a base for an evolutionary campaign on the notched seed-shrimps in the Caribbean.

There are other researchers investigating notched seed-shrimp bioluminescence in the USA. In the early 1980s, Jim Morin, then of the University of California, Los Angeles, went in search of bioluminescence on the reefs of the Caribbean Sea. What he found was unexpected. There were the usual starfish and worms glowing as they roamed sloth-like over the seafloor. But in the open sea above them were luminous flashes that rivalled those of fireflies on land for their spectacular exhibitions, appearing like a firework display. The fireflies of the sea were in fact notched seed-shrimps. Later, Jim Morin was joined by Anne Cohen of the Los Angeles County Museum of Natural History, who had been rearing notched seed-shrimps in her lab. Considerable documentation and analysis of the Caribbean bioluminescence followed.

It became evident that different patterns of flashes were being produced in the Caribbean waters. Soon after sunset, blue lights would be flashed in the water column, one swiftly following another, to create specific patterns like constellations in the sky. About fifty different patterns were identified in total. A sequence of about ten flashes would take a few seconds to complete, and the eye would always be drawn in the direction of the pattern. Sometimes the flashes would move vertically upwards in the water, sometimes directly downwards. Some flashes would move horizontally, others at an angle, while sometimes single flashes would be replaced by groups of flashes, all moving in unison to create a new pattern. Within these sequences, individual flashes could be evenly spaced or become increasingly closer to their neighbours. All quite spectacular.

The notched seed-shrimp maestros were captured in nets in mid-performance. They were all males, but were being tailed by female notched seed-shrimps. The Caribbean males would emerge from the sand, swim into the open water and flash their lights. These seductive dances would catch the eyes of females, luring them too into the water column. From here on they would be uncontrollably attracted towards the males, and presumably all would be in the mood for mating. Although mating could not be observed with the low magnification cameras employed underwater, evidence was found to suggest that these flash patterns really were courtship rituals, like the iridescent display of the *Skogsbergia* species in Australia.

The males producing the horizontal pattern, and the females attracted by this pattern, all belonged to the same species. Similarly, the males and females associated with the angled pattern all belonged to the same species, a different one from that of horizontal persuasion. And so the story continued, until some fifty different species were found to match around fifty different patterns. In the Caribbean, it seemed that notched seed-shrimps had evolved a nice strategy for mate recognition and courtship – it *had* to be really efficient to outweigh the disadvantages inherent in making oneself so conspicuous to predators. This was a strategy where many species could be packed into a restricted environment and still easily recognise and mate with their own kind. Mistakes, potentially as costly to a species' hopes of survival in the long term as they are embarrassing in the short term, were minimised. This brings us to the subject of evolution.

Lou Kornicker of the Smithsonian Institution in Washington, DC, had produced taxonomic publications the size of telephone directories on lightweight seed-shrimps, including notched seed-shrimps. His work provided a reliable database of body parts and the variety of forms of notched seed-shrimps. And an evolutionary tree was inferred at last.

The global view – evolution of all notched seed-shrimps

The evolution of the Caribbean species was analysed in further detail. It emerged that similar looking flash patterns of bioluminescence belonged to species that were closely related. So the evolution of flash patterns was not haphazard, but rather orderly, in a stepwise manner. A disordered evolution would have implied the patterns were adaptive: adapted to the specific environment of a species. But a gradual evolution inferred the flash patterns were evolving in synchronisation with the species themselves. So what can be learnt from all of this? Before advancing further with this line of thought, notched seed-shrimp iridescence should be reconsidered.

The evolutionary tree of notched seed-shrimps revealed a trend – bioluminescence appeared only and always in one half of the tree. All bioluminescent species were related – bioluminescence had evolved just once in notched seed-shrimps, and was retained in all descendants of the forebear. At another level, the bioluminescent half of the tree could be further divided into those species that produced patterns of flashes, and those that flashed only to avoid predation. At the beginning of the complete tree stood the baked bean, and bioluminescence evolved a few branches later. A broader investigation of diffraction gratings revealed that the bioluminescent flashing species all possessed fairly similar and rather rudimentary halophores like those of the baked bean. So halophores, and consequently iridescence, had not been evolving within the bioluminescent flashing branches of the tree. Meanwhile, the remainder of the notched seed-shrimp tree of life was telling a different story.

We have learnt that the diffraction gratings of notched seed-shrimps can be ordered into a neat sequence. This sequence becomes increasingly clear when bioluminescent species are disregarded – the bioluminescence species were clumped together at the start of the sequence. It so happens that the order of species within the sequence of iridescence matches precisely the order of the species inferred from the evolutionary tree, from those that derived earliest from the seed-shrimp ancestors, to the most recently derived. So the members of the non-bioluminescent half of the evolutionary tree have been gaining increasingly efficient diffraction gratings and, consequently, light displays. At the very top of this iridescent half of the tree was *Skogsbergia*, the movie star.

Considering that bioluminescent flash patterns and iridescent displays are employed for mating purposes, they surely now have implications for evolution. If genetic mutations occur when an individual is conceived, the diffraction gratings of an offspring may be different from those of its parents. If the mutation is somehow advantageous, such as being a more efficient signal for mating, it can be retained within the future evolutionary line. A more efficient signal for mating, in the case of the notched seed-shrimps, would be a more complex pattern of bioluminescent light or a brighter, or bluer, iridescence. Blue light travels best or furthest through sea water, with green not far behind. If the new design of signal mutates further throughout the future evolutionary line, the signal of the future can become unrecognisable from the original, ancestral signal. Eventually a point is reached where the ancestral forms, which have continued to reproduce without signal mutation, can no longer recognise the 'future' signal. Considering we are talking about a code for courtship, the ancestral forms can no longer mate with the contemporary signallers. A new species has evolved. The new species would appear as the most derived on the evolutionary tree, at the tip of the branches.

An analogy to this story could be found among human beings. Humans adorn themselves with clothes, scent, jewellery or body art to attract the opposite sex. Different races of humans decorate themselves to different extremes, so much so that a female of one race will not necessarily attract the male of another, or vice versa. Consider those Amazonian men with plates inserted in their lower lips. European races, for instance, probably would not find this particularly alluring, and so Europeans and Amazonians do not interbreed. This keeps the races separate, and thus is analogous to the different species of seed-shrimps in our story. But imagine a new trend emerging in the Amazon where, in one village, it was no longer considered attractive to possess a plate in one's lip, but rather a tattoo on one's face. Before long a new race would have emerged following the incompatibility of plate-bearing and tattoo-wearing individuals, based on courtship display. The two races in the Amazon are now as divorced from each other as they are from Europeans, although still more closely related to each other on the tree of races. It should be made clear that this is not a case of evolution, but human invention. Returning to evolution, the notched seed-shrimp story can continue from here, with new species evolving that bear more attractive or flamboyant costumes.

The point of this whole story, and this chapter so far, is that notched seed-shrimps appear to have been evolving to become increasingly well adapted to light. The very first notched seed-shrimps of 350 million years ago are represented today by the living fossil, the baked bean, with its primitive form of diffraction gratings. In subsequent evolution, light became something to which the notched seed-shrimps adapted strongly. Light has imposed a momentous selection pressure throughout their evolution. In fact their adaptation to light may even explain the evolution of the notched seed-shrimps, via the changes that took place in their courtship displays. This is nice to know. It is one thing to determine an evolutionary tree, but something else altogether to explain it. Here we can explain why different species of notched seed-shrimps were evolving. But the important message for this book is that light can have a powerful influence on evolution. And this does not apply only to notched seed-shrimps, as will be demonstrated after an epilogue to the seed-shrimp.

The strong adaptation to light has been a hugely successful strategy for the notched seed-shrimps. The fossil record suggests that 350 million years ago notched seed-shrimps were rare. The SEAS project revealed that today they are the commonest multicelled animal group on the Australian continental shelf at least. An evolutionary success story for these seed-shrimps, with a happy ending . . . so far anyway. Evolution continues.

Natural diffraction gratings

Another important conclusion to emerge from this study of seed-shrimps was that diffraction gratings do exist in nature. This finding itself became the foundation for another project – to unearth any other diffraction gratings that lay hidden within the animal kingdom. Confidence in a positive result was high because now it was known what to look for. And indeed further cases emerged. But more unexpectedly, another link between diffractive structures and evolution appeared, in the case of the upside-down fly.

Diffraction gratings were found within a range of invertebrate animals, from the hairs of worms to the wings of flies. In fact the bristle worms are particularly well endowed with diffraction gratings, and reveal a variety of diffractive forms. This finding will become important later in the following chapter.

In addition to strict diffraction gratings, similar structures were discovered which also cause sunlight to diffract, but this time the light reflected would appear a metallic white, or silver, in colour. This resulted from diffraction gratings running in a variety of directions, where their reflected spectra would overlap. Sunlight would be split into its spectrum, which would then be reconstructed. This was comparable to Newton's famous 'two-prism' experiment, where one prism cleaved sunlight into its component colours, while another was positioned to

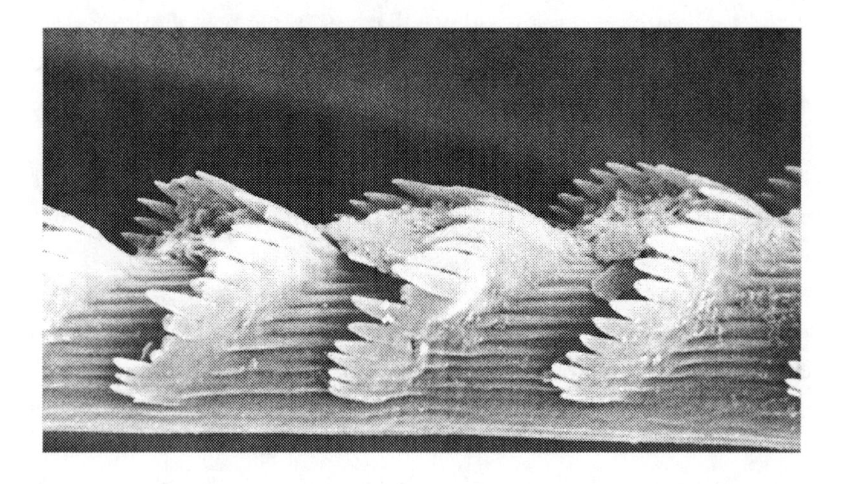

Figure 5.5 Electron micrograph of a hair from *Lobochesis longiseta*, a bristle worm. The ridges are spaced about one micron apart, forming a diffraction grating that causes a spectral effect.

recombine the colours. After passing through the two prisms, normal sunlight resumed. The mechanism of reflection in the newly discovered diffractive structures was essentially the same as for scattering, where microscopic particles reflect all wavelengths in sunlight equally in all directions. The fibres in the paper of this book perform this task. There was, however, an angular attribute to the newly discovered structures – white light could be reflected in just one direction. This equated to a very strong reflection if one happened to be looking from this direction. And the most impressive effect of all belonged to the upside-down flies.

Australia's upside-down flies

The upside-down flies fall within a group of small flies described by David McAlpine of the Australian Museum. David McAlpine also noticed similarities between and peculiarities in the behaviour of species from this group. Some plants have long, vertically upright leaves. These leaves provide a home for the 'McAlpine flies', which often rest on the leaves in a group, bodies oriented vertically. The gathering of the flies is an act of safety – there is, after all, much to be said for safety in numbers. Also the possibility of reproduction is enhanced if potential mates are close by and easy to find.

The flies involved in this story belong to many species, and collectively live in Africa, Madagascar, South-East Asia and Australia; the ancestral species became divided as the continents separated millions of years ago. In fact the ancestral fly *is* known: one specimen has been found preserved in amber in wonderful condition.

I borrowed the amber specimen from a museum in Göttingen, Germany. The amber has been fashioned into a neat block, about a centimetre square and a few millimetres deep, and is mounted on a glass microscope slide. Inside the amber are two flies, one the size of a large mosquito with big, perfectly preserved eyes, and a smaller specimen, which is the one of interest here. The smaller fly was described by the German biologist Willi Hennig, a man rather more famous for his development of a phylogenetic method – the main tool used to study evolution today. Unfortunately, the fly is orientated in a most inconvenient way. Along with inconsistencies in the amber, it can be seen only in a limited and distorted view, so it is not easy to say whether or not this ancestral specimen possesses reflective patches. To make matters worse, amber would affect the optical properties of many reflector types, such as diffraction gratings. We would need to see this fly in air, not in amber. And the rarity of this specimen has resulted in a ban on any potentially destructive handling, so an informative dissection is out of the question.

The living relatives of this amber specimen, however, do possess light reflectors – diffractive structures based on a system of hairs that appear silver. The hairs come in different shapes and sizes, and can be aligned differently, although always spaced evenly. The specific, microscopic characters determine the optical properties of the complete reflector, which vary from species to species. And a pattern emerged from a study of this variation.

The fly in amber evolved along two separate paths. Like the history of notched seed-shrimps, the evolutionary tree originating from the fly in amber can be divided into two halves. On one side we have the right-way-up flies, and on the other side the upside-downs. But all have one thing in common – they reflect silver light upwards, towards the sky. This reflection probably acts as a signpost to other flies in the vicinity, to invite them to a gathering.

The right-way-up flies, orientated vertically on their host leaves, go about their business with heads facing the sky. They live in South-East Asia and Australia. Those ('primitive') species of right-way-up flies with the oldest ancestors possess very inefficient reflectors, positioned between the eyes so that sunlight can be reflected back towards the sky. This reflective patch must have proved rather useful to the fly. It was not only passed on to the next species in the evolutionary line, but it was also improved upon. The physics of the reflector became more efficient, and, consequently, its visual effect became more striking. This trend can be traced through the entire evolutionary line of the right-way-up flies. The next species to evolve not only improved upon the physics of the reflector again, it also sprouted more reflectors over its body. The additional reflectors appeared only on other sky-facing parts of the fly, such as the front parts of the first pair of legs. A greatly speeded up film of evolution through geological time would reveal reflectors blooming from increasingly more parts of the upward-facing body. Furthermore, the reflections would appear increasingly brighter as the optical properties kept on improving via evolution. And exactly the same was happening, independently, in the other half of the evolutionary tree.

The upside-down-flies live in Africa, Madagascar and Australia. They are so called because a strange thing happened at the beginning of their history. As the ancestor represented in amber evolved in this half of the evolutionary tree, it turned upside-down. Still living on vertical leaves, and still with its body orientated vertically, it turned through 180° – and never looked back. The upside-down flies all face the ground, so that their rear ends point towards the sky. This could be explained by their occupation of slightly different plant species – plants with predatory spiders lurking near their leaf bases. So an upside-down fly could keep a lookout for dangers from below. The upside-down flies continued to aggregate, and probably also employed reflectors to call to their friends. So how could they signal towards the sky when they are facing downwards? They simply 'moved' their reflectors so that they faced the other way.

The upside-down flies have reflectors on the backward-facing parts of their bodies. And they evolved almost in tandem with their right-way-up counterparts. Again the reflective patches increased both in abundance and efficiency throughout the evolution of the upside-down group. There were, however, differences in the designs of reflectors between the right-way-up and upside-down flies. In fact it is the most recently evolved upside-down fly that owns the most efficient reflector. This champion upside-down fly lives, of course, in Australia, and from rudimentary beginnings it has evolved a type of reflector never before seen in the world of optics, let alone biology. This could even have been applied to human optical devices. But it is the evolutionary tale that is relevant to this book.

The group of McAlpine flies, like the notched seed-shrimps, has evolved with light as a major stimulus. We may be so bold as to say that light has driven the evolution of this group. And this second example of the influence of light on evolution is not the last. Back in the sea, light has been known to insert its influence on the evolution of a group of crabs.

From sound to light

The snapping, or pistol, shrimp has one small and one large crab-like claw. The large claw is the pistol, which fires an underwater bullet of sound so loud that it can be detected by passing submarines. In fact it can even interrupt their sonar. Sound can have its drawbacks because it is an omnidirectional signal – it is, as the word suggests, sent out in every direction. So not only does one reach a target organism, but also every other organism in the vicinity.

Another crustacean making sounds in the sea is the oval (swimming) crab. Although spending most of its time on the sea floor, it is equipped with swimming paddles on its rear legs to propel it through the water whenever required. Individuals aggregate on the sea floor to form species groups, which are highly aggressive towards each other.

There are many species of oval crabs and the ancestral type, known from fossils, made sounds in prehistoric times. This ancestor possessed a file and pick that scraped together to make trademark music audible in ancient seas. This was probably the oval crab's means of attracting its own species for aggregation. And it was successful because those sounds can still be heard today, made by descendants of the ancestral species. In fact about half of the oval crab species living today employ a similar instrument. The oval crabs, nonetheless, have greatly increased their diversity by succumbing to a selection pressure other than that for sound production – sunlight.

The ancestral oval crab and the living musical species all have strong shells. Their shells are strong because they are composed of a stack of thin layers. In fact a cross section of their shells appears like a multi-layer reflector, except that the layers are too thick to cause a reflection. Still, a stack of layers is stronger, tougher and more resistant to cracking than a continuous slab of the same material. Think of pieces of wood used in DIY that are composed of thin layers glued together, for instance; they are both strong and effective.

Although one group of oval crabs continued with their music-making, another group gradually lost the ability to make sounds, while progressively acquiring the ability to reflect light. Throughout the evolution of this colourful group, the picks and files gradually diminished until they vanished completely. But early on in the evolution of this group, a change took place in their shells – the composite layers became thinner and, to maintain the overall thickness of the shell wall, more numerous. While retaining their strength characteristics, the layers had formed into multilayer reflectors. Shells began to appear iridescent.

The first oval crab species to evolve iridescence retained some ability to make sounds, which was probably useful because only a small area of its shell had become iridescent. It was a shy flasher. But then the floodgates opened. The next species to evolve contributed a greater spectrum to the oceans – it was more extensively clothed in iridescence. And so on until the most spectacular marine animal of all had arrived on Earth – the majestic iridescent crab. This is a large crab, with a shell the size of a grapefruit, that gleams with brilliant iridescence from every part of its body – shell, legs and claws. Imagine a crab made of the most spectacular opal. There would have to be great advantages to having such a bright attire, because the disadvantages are obvious, particularly the way the crab continually advertises its presence to predatory fish. Where seed-shrimps succeeded in concealing their iridescence when it was not required, oval crabs failed. But the iridescent advertising of the majestic iridescent crab is not as pointed as would at first appear to be the case because it has a trick up its sleeve – in its natural environment it can make itself invisible. Here lies an advantage of an iridescent signal – it is directional. Compare the explosion of a pistol to the flash of a torch in a bright environment. Unlike the explosion, the torchlight can only be detected when one looks directly at it. Either way, there really must be advantages in appearing brightly coloured, because the oval crabs that evolved iridescence also devolved their sound production. Time does tell. But these advantages could be confined to certain areas of the globe – the areas where the colourful oval crabs live. Maybe predatory fish have 'bigger ears' in these areas, so it is best to keep quiet.

Again, the relevant conclusion to be drawn for the purposes of this chapter is that light has played a major role in the evolution of an animal group. Sunlight could be considered the driving force for the evolution of the iridescent half of the oval crab tree. And the momentum of evolution in the direction of the sunlight selection pressure never slackened.

The list continues

I have dealt only with structural colours in this chapter because these can be represented by mathematical equations and granted efficiency values rather easily. But changes in pigments are also known to occur throughout evolution. Nudibranchs, or sea slugs – marine snails that have lost their shells (through evolution, of course) – demonstrate just how spectacular a pigment can be. Some of the most memorable underwater photographs seen in the coffee-table books on marine life are of sea slugs. But the taxonomy of sea slugs is problematic. Once their pigments have broken down in preservative, and their colours have faded completely, many of them look extremely similar. It is their colours that separate them into species without the aid of dissections or genetic analyses. Their unmistakable colours provide warnings to predators. Different species have different predators, and their colours have evolved to suit. As predator vision changes or evolves, so do the sea slug colours. Hence light is a major selection pressure to the evolution of sea slugs.

There are many other examples of evolution driven by light. Light is not only a governing factor of animal behaviour at any one point in time, such as today, but is equally important in the evolution from today's ecosystem to that of the geological tomorrow. Light not only exposes an animal as conspicuous or camouflaged today, but can also drive the evolution of animals in the future. As inferred in Chapter 3, if an animal is not adapted to the light in its environment, it will not survive. And light is an exception among the stimuli because in most environments it is always there. One cannot ignore light. But equally important to this book is the issue of evolutionary dynamics. It is one thing to know *what* happens during the course of evolution, or the design of the evolutionary tree, but something altogether different to explain *why* it happens. In this chapter it has been demonstrated that adaptation to light can be the *why* of evolution. And the next questions to emerge are, of course, 'Was light a selection pressure at the time of the Cambrian explosion?' and, if so, 'How strong a selection pressure was it compared with others?'

The disparate subjects of colour and animal evolution have emerged as compatible. This chapter signals the dawn of a relationship that will mature as subsequent chapters unfold. The perseverance with colour is beginning to pay off, as clues begin to gather towards solving the Cambrian enigma that this book is attempting to understand. But these are still early days in the Cambrian trial, and evidence also needs to be sought from other avenues.

So far we have examined colour in living animals and predicted the course of colour evolution. But is there also real evidence of colour in the past? Can we return now to the fossils and hope to unearth their true colours? If so, this may be a step towards finding the answers for the above questions about Cambrian light. Armed with our understanding of colour today, it is certainly worth a closer look at what was described in Chapter 2 as a void in palaeontology. In Chapter 6 I will attempt to start filling that void.

6

Colour in the Cambrian?

All species still glow in their original, almost fantastic array of colours

HERBERT LUTZ, German biologist, on the colour of 49-million-year-old jewel beetles from Messel, Germany

Today the Museum of Antiquities in Leiden in the Netherlands, houses a statue of the Egyptian god Osiris. This statue is about a foot tall with well-preserved features, and also a fair amount of its original paint – it has seen little sunlight, being mostly preserved within a tomb. Here, Osiris has a blue-green face and wears a red skirt. And another obvious feature of this statue is that it is hollow . . . but why? Without the preserved colour this question would remain unanswered. Numerous statues of Osiris have been excavated but the hollow inside and colouration make this particular representation different.

The interpretation of hieroglyphics, and the preservation of yet more pigments in the form of ancient Egyptian scripts, inform us that blue-green was the colour used to represent the afterlife and red was used for festivity. So now we can interpret this statue of Osiris as being a celebration of the afterlife. From this, and the knowledge that hollow Egyptian figures were filled with papyrus manuscripts, we can infer that our statue once contained a copy of the Egyptian Book of the Dead.

The ancient Egyptians were, in fact, skilled artists. They used colour to represent personality and status, but they knew it would fade with time. Consequently much of their art was sculpted and then painted, so that at least the physical sculpture would remain long after their death (as was their intention). But they also had gold leaf at their disposal. The cause of the gold effect in this case lies somewhere between a pigment and a structural colour. Gold leaf is a thin layer of metal that reflects a beam of sunlight in a single direction, like a mirror. It reflects all the wavelengths in sunlight except blue, all of which add up to gold. As a physical structure it outlasts the pigments of ordinary paint through time. So gold leaf was used on many Egyptian statues, since the Egyptians were conscious of the short-term prospects of their pigments. And gold leaf is indeed evident in numerous Egyptian artefacts today, as in another statue of Osiris housed in Leiden. The gold in this case is symbolic of eminence.

Chapter 3 demonstrated that colour alone tells us about where and how an animal lives today. Considering the information acquired from the colour of the pigments in the Egyptian statue of Osiris, a question relevant to this chapter now begins to form: 'Can we bring Cambrian fossils to life in the same manner?' The excellent preservation of gold leaf in the statue of Osiris signals hope of unearthing structural colours of geologically ancient times.

We know that animal body shapes and forms were as complex in the Cambrian as they are today, so perhaps we can also expect Cambrian animals to have been sophisticated in terms of their colour. But we have learnt not simply to predict colour based on animals today. We must find traces of the *original* colours themselves in ancient, extinct animals. And the best place to look is in those fossils that have been preserved under the most favourable conditions. Work in this field is already underway.

Trilobites that lived 500 million years ago, just after the Cambrian, have been found with signs of pink colouration, not something that is easily explained given the type of rock in which they were preserved. It is therefore believed that these randomly arranged pink pigment granules are remnants of a colour that once covered the entire trilobite. That would be interesting. Below the very surface waters, and in the environment inhabited by these trilobites, red light does not exist. Here, pink becomes grey and blends well into the background. So these trilobites may have been coloured for camouflage. But few experiments have been conducted in this case, and so speculation must end there. And this case of trilobite-pink also represents the end of the road for ancient pigments. Unfortunately, pigments, and also bioluminescent organs, do not take us back to the Cambrian, and so can be of little use to the subject of this chapter. But structural colours are another matter altogether. Could *they* tell us anything about colour in the Cambrian?

As outlined in Chapters 3 to 5, physical devices that cause 'structural' colours are a significant means of light display today. Like pigments, structural colours rely on a source of incoming light, usually in the form of sunlight, from which certain wavelengths, or 'colours', are reflected.

Structures can be preserved in the fossil record – at least their shapes and sizes can be, even if the original materials become altered or replaced. Fossils themselves, whether the whole bodies of trilobites or the bones of dinosaurs, are indeed structures. Although on a much smaller scale, it is therefore not surprising that structures responsible for colour can also be preserved within fine sediment – these are, after all, just structures. Obviously micron-sized reflectors could not be preserved in sediment of 1 millimetre sand grains – apart from the obvious physical problems they would be consumed by the bacteria infilling the spaces between grains. This was the reason why we cannot find minute sensory detectors in the Australian Ediacaran (Precambrian) fossils. Shapes of the entire animals can be seen with the naked eye, but under a microscope nothing more than piles of sand grains can be distinguished. Similarly, the chemical components in the embryonic rock must be right to replace organic parts. But there is certainly greater potential for structures to be recorded in the fossil record than for pigments. And if the conditions are right, theoretically there is no lower limit to the size of a structure that can be preserved as a fossil.

Before moving directly to the Cambrian fossils themselves, we should take a look at the methods at our disposal for unearthing ancient hues. We should be aware of the variety of structural colours that may be preserved along with some of the pitfalls one may encounter along the way to the Cambrian.

Ammonites – multilayer reflectors and modifications

We know that multilayer reflectors are the most widespread cause of structural colours in animals today. Like pigments, these occur within the bodies of animals, below the surface. Again, the scanning electron microscope is not appropriate here because it can only scan surfaces. So to search for multilayer reflectors, we must look at thin sections of fossil skin or shell – the outer layers of an animal. Some years ago I tried exactly this, using ammonites and ancient beetles as my guinea pigs.

Ammonites are among the few groups of animals whose original, transparent, thin layers have survived in fossils, and colours radiate from some of them today as they may have appeared millions of years ago. But this cannot be assumed for every case of iridescent fossils. There are warnings to heed from opal – all that glitters may not be old, or at least not as old as the animals that have been fossilised.

In Chapter 5 I described my discovery of structural colour in seed-shrimps, almost the first structural colour known in seed-shrimps. A couple of years earlier, while sorting through a large sample of small crustaceans, I had noticed a single flash of colour from one seed-shrimp. There were many other individuals of this species, and all were quite transparent, but while I moved the sample one individual was flashing red one minute, and green and blue the next.

The seed-shrimp was the size of a tomato seed, and the source of the colour much smaller, but it was large enough to be identified under the microscope. The identification also solved the problem of why only one individual should reveal colours. The source of the colour was not a feature of the animal itself, but a tiny opal, and the seed-shrimp had eaten it. The opal lay in the stomach of the transparent animal.

Opal is a form of silica dioxide. It is made up of tiny spheres, around half the wavelength of light in diameter. It reflects light in a complex manner, which has only recently been understood by optical physicists. But it is the physical nature of the structure that provides the optical effect rather than a chemical pigment, and so opal is said to produce structural colours. In fact the bright, iridescent effect of opals is similar to that of the seed-shrimp diffraction gratings.

The original chemicals that make up fossils, at whatever stage in the fossilisation process, can be replaced by other chemicals. Sometimes, the replacement chemicals can be silica dioxide and water, in which case opal is formed in the mould that is the fossil. At Lightning Ridge in Australia, opal miners often excavate dinosaur bones and teeth, and the parts of other animals, which display the characteristic iridescence of opal. These fossils are so well known that most palaeontologists think of them when we mention 'colour in fossils'. But unfortunately this adds no evidence to the original colours of ancient life – opal has nothing to do with living animals (other than that single seed-shrimp).

Ammonites are the shells of ammonoids, those long-extinct molluscs related to squid as described in Chapter 2. Some ammonites appear coloured, but like opal their hues are non-biological. Particularly striking for their visual effect are the ammonites from Alberta, Canada, which flash spectacular colours as their rocks are cracked open.

In view of the Canadian Rockies lies the small town of Magrath, and the familiar wheat fields and ranches of the Canadian prairies. Seventy-one million years ago, this land was beneath a sea which stretched from the Gulf of Mexico to the Arctic Ocean. And in this sea lived ammonoids – lots of them, ranging from the size of a compact disc to that of a car tyre. Today, one particular ranch near Magrath, of about 800 hectares, is different from the others. Its foundations contain ammonites.

These ammonites were first covered not with sand but with ash from the huge volcanic eruptions – which played a part in the creation of the Rocky Mountains. The ammonites became sealed in a waterproof layer of shale, but this did not prevent quartz, copper and iron from the volcanic ash infiltrating the shells. During the Ice Age, a layer of ice close to 2 kilometres thick covered the region. The weight of this ice served to compress the ammonites and their component chemicals, and 'Ammolite' was formed.

Ammolite (and Korite) are names given to a semi-precious gemstone that partly constitutes the Magrath ammonites. In 1981 enough high-quality Ammolite was discovered to make mining commercially viable. But their equally commercial bright colours are the result of preservation, the compacting of the shell layer that *may* have possessed some iridescent properties to begin with. Many shells today have an iridescent layer, containing a multilayer reflector called the nacreous layer. We suspect the Magrath ammonites might also have contained a nacreous layer because other ammonites have been found in a more natural state, also with iridescence.

In Wootton Bassett in Wiltshire, England, ammonites literally pop up out of the ground, for 20 metres below a spring, Jurassic clay in the form of grey mud oozes to the surface in a sort of mud volcano, bringing with it Jurassic ammonites hitching a ride in the eruption. Although 180 million years old, these ammonites are also iridescent, but they are different from those found in Magrath. The Wiltshire ammonites are pristine fossils, unaltered since their initial preservation. Inside the shells are some original organic ligaments, but they also retain their aragonite, a calcium-based mineral and a component of their original shells. It is this aragonite, within the nacreous layer of the shell, which is responsible for the iridescence. Aragonite forms thin layers, each a quarter of the wavelength of light in thickness and all separated by a similar distance. Consequently, the nacreous layer is a multilayer reflector, like those found in metallic beetles and shells today. But as explained in Chapter 3, multiple layers can also provide structural strength, and when strength is the adaptive function, the incidental iridescence is nullified by an opaque, outer covering. Iridescence is a powerful effect, and redundant iridescence would be simply too dangerous to project recklessly into the environment. A camouflaged soldier could not smoke a cigarette in the evening, especially if the light from the cigarette was not also being used as torchlight. So although iridescence is quite eye-catching in these ammonites today, and in specimens from other parts of the world, in the Jurassic the story could have been quite different. The prehistoric seas could have been spared ammonoid iridescence by a dark outer layer of their shells, a layer that has not been preserved. Ammonoids will pop up again later in this book, but now we should consider those fossils whose original colours are displayed today just as they were in environments some fifty million years ago.

The Messel beetles – original multilayer reflectors

There is one particular quarry in Messel, near Frankfurt in Germany, that reveals extraordinarily preserved, articulated skeletons of vertebrates, around fifty million years old, surrounded by complete outlines of their bodies. This quarry also contains insect exoskeletons like no other fossil site – chitin, the primary component of arthropod shells, has been preserved there.

Today the bowl-shaped crater at Messel is fenced off and closely guarded. It is now generally accepted that something special occurred here, but this was not always the case. When the mining that originally created the crater came to an end in the 1960s, the intention was to infill the site with garbage. Then it was that fossils found when quarrying first began were brought to public attention. Almost immediately the United Nations declared Messel a World Heritage site.

Forty-nine million years ago, after the mass extinction that killed off the dinosaurs, Europe was an island and the Messel site lay at the bottom of a lake. Today the rock in the quarry is still damp – it is 40 per cent water. But when the layers of thin sediment are cracked open, they sometimes reveal a little more. Fossils of entire animals, from bats to crocodiles, have been exposed. Preservation is so good in this oil shale that Messel palaeontologists tend to feel more like zoologists. But when the fossils are exposed to air, they must be immediately stored in water, for the rock crumbles if it dries.

Structures such as the feathers of birds have been preserved at Messel as if they had only moments earlier fallen from the sky, but in my biased opinion the greatest treasures of all at Messel – and justification alone for the high security – are the metallic-coloured beetles. Their optical effects are extraordinary. Stag beetles reflect the shimmering blues and greens they displayed while alive. As the shale containing a jewel beetle is broken, the sight of 49-million-year-old iridescent yellows and reds is revealed. And so the list of beetles and colours continues.

Sometime in 1997 I received a parcel from Germany from the palaeontologist Stephan Schaal, a man whose name is almost synonymous with that of Messel. As I had hoped, it contained Messel beetles from a recent excavation. The fossils were stored in water, and the wing cases shimmered with violet, blue and green. Since colour in animals was at the centre of my research, the first question to cross my mind was, 'What is causing this colour?' The age of the fossils simply did not register – the beetles looked like zoological museum specimens recently collected from a rainforest expedition. After all, 49 million years is a long period to comprehend or time travel in one's mind.

To answer my question, I turned to electron microscopy. Small sections of a blue beetle exoskeleton were treated in two different ways. One section was critical point dried – that is, it was dried out in a controlled manner to prevent shrinkage. Although it had retained its structure, the dried section had lost its colour. It had become transparent. To examine the structure in the scanning electron microscope, it was first coated with gold. Then, at 10,000 times magnification, thin layers became evident, with upper layers only partly overlapping the lower layers. The layers were smooth, and there was no sign of a diffraction grating or structures that could cause the scattering of light. But to confirm that this was a multilayer reflector, transmission electron micrographs were needed.

One of the beetle sections was embedded in resin, stained and sliced so thin that it was not visible edge on. Placed on a minute metal grid to provide support, the specimen was imaged in an electron beam. A multilayer reflector was revealed.

To be doubly sure, the dimensions of the reflector were measured and fed into a computer program which re-created the stack of thin layers and predicted the colour of the light reflected in sunlight at 90° to the surface. The predicted colour was blue. The actual colour I saw was blue. Hence the cause of the colour of Messel beetles *was* a multilayer reflector. The reason the colour had disappeared from the dried specimen also could have been predicted. It emerged that one of the two layer types in the reflector consisted of water – when the water disappeared, so did the colour.

Several specimens of the same beetle species have been found at Messel, and all display exactly the same colours. So we can be confident that 49 million years ago beetles were gracing Europe with spectacular iridescence – last seen flashing when the dead beetles were washed into the Messel lake by floodwater, and were sinking into the depths of history. Re-opening the history book, we learn that light must have been a powerful stimulus to animal behaviour even then. But how far back can structural colours help us take this philosophy?

Fossils of the Burgess Shale – diffraction gratings

In 1966 Kenneth Towe and Charles Harper, palaeontologists at the Smithsonian Institution, published a paper describing the cause of iridescence in 420-million-year-old lamp shells. They found tubular aragonite crystals arranged in layers, with dimensions in the region of the wavelength of light. A layer of juxtaposed tubes may create a diffraction grating on the outside, but a stack of thin layers can also form a multilayer reflector. The lamp shells appeared with a rather faint iridescence or pearly lustre like that of some shells today. Towe and Harper suggested the cause of this optical effect was a combined grating-multilayer structure, and attributed the faintness to variations in spacings, or a degree of randomness in the structure. Further work may be required to confirm these conclusions, but we cannot say for certain that these colours were sparkling in 420-million-year-old waters. Again, considering shells today, the lamp shell iridescence may have been precluded by an opaque outer layer, a layer that was not preserved in the fossil.

True diffraction gratings are well known to physicists, but before my search, prompted by their discovery in seed-shrimps, they were unknown in nature. Then diffraction gratings began to appear in one animal after another. First there was a lobster found off Hawaii, then a type of shrimp from New Caledonia, again in the Pacific Ocean. But the Indian Ocean was hiding similar treasures, not only within its crustaceans but in bristle worms, comb jellies, jellyfish and peanut worms. Eventually it was discovered that the entire globe contained a vast array of species, from many animal phyla, loaded with diffraction gratings. The world, it turned out, was even more colourful than we had believed it to be, albeit that the newly revealed iridescence was often concealed from view for most of the time.

Part of my work on seed-shrimp iridescence described in Chapter 5 was carried out at the National Museum of Natural History of the Smithsonian Institution. Originally I had found diffraction gratings in some seed-shrimps from Australia and needed to examine as many other species as possible. The world's expert on this group of animals is Louis Kornicker at the Smithsonian, and it is no coincidence that the best collection of seed-shrimps is found there, too. So it was only natural that I should apply for funding to work in Washington. My application was successful and in 1995 I began working on the Smithsonian collection.

As mentioned in Chapter 1, the Smithsonian also houses probably the best and certainly the most important collection of Burgess Shale fossils anywhere in the world. Now that is a coincidence. The Smithsonian was the home of Charles Doolittle Walcott, who discovered the first Burgess Shale fossils. But other than a general fascination in this 'wonderful life' evident among all zoologists, I had no specific interest in the fossils themselves.

Taking a break from work at the Smithsonian, one is spoilt for choice for things to do. Within a few blocks of each other on a single avenue there are several national museums and art galleries. But there was also the Museum of Natural History, and during one late afternoon break I found myself wandering around the fossil galleries.

I discovered a small but excellent exhibit on the Burgess Shale nestling between larger skeletons. This exhibit was worthy of its space because the fossils displayed were complete and detailed examples of the range of life forms for which, in addition to its age, the Burgess Shale was famous. The specimens were also those collected mainly by Walcott in the early 1900s.

Next to each fossil in the exhibit were black and white illustrations showing reconstructions of the animals when they were alive. The drawings were very detailed and really helped one to visualise the *living* creatures. But the level of detail included something of interest to me specifically. On some reconstructions there was a hint of something quite amazing. On the reconstructed armoured parts of *Hallucigenia* and *Wiwaxia* were fine parallel lines. And fine parallel lines were the reason I had come to Washington in the first place.

The day before, I had visited the aviation and space museum which housed some aeroplanes from the 1950s, each with multiple propellers and corrugated wings and fuselage. The corrugations served to increase the strength of the metal structures. Later I was to encounter similar corrugations used to increase structural strength – but in the leaves of a Rocky Mountain plant on my expedition to the Burgess Shale quarry. These leaves were thin and would have collapsed were it not for their corrugated form. This was important to bear in mind. Narrow striations on the Burgess Shale fossils could represent a finely corrugated surface to make them stronger. But it got me thinking. The same rules apply to animals today, although if the striations meet certain size criteria, they cause iridescence – they become diffraction gratings.

Reference to diffraction gratings insinuates microscopically fine corrugations, where a distance approaching the wavelength of light separates neighbouring ridges. Such structures cannot be drawn as lines on paper. No pen is that sharp or precise, and we would not be able to see the lines with the naked eye anyway. But, as I have said, this got me thinking. Maybe the lines figured in the animal reconstructions were merely representatives of diffraction gratings. Fossil preservation is never uniform, and perhaps only some ridges of a grating had been preserved. Then again, maybe the lines figured were complete and did serve to provide strength. If this were the case, the parallel lines illustrated in the Smithsonian exhibit would add nothing to the topic of colour in fossils. But they changed the direction of my thoughts. If animals possess diffraction gratings today, maybe they did so in the past too.

The morning after my first Burgess Shale viewing, I submitted a request to examine the original Cambrian finds. Access to the Smithsonian fossils, and those at Harvard University, was granted, following support from Simon Conway Morris of Cambridge University, Doug Erwin at the Smithsonian and Frederick Collier at Harvard. Then there were some specimens to examine at the Australian Museum back in Sydney. To begin with I employed the most powerful light microscopes available at the Smithsonian Institution. I had not realised that the compact disc case I was using to orientate the specimens under the microscope was titled 'Handel's Water Music' – quite appropriate, as certain onlookers remarked. But it worked. I placed the fossils so they could be viewed from different angles, and structures I had not noticed before became evident. Then I knew exactly what to look at, and the work became serious. I took the fossils to the various underground rooms of different institutions, where vibrations and magnetic fields that could interfere with more powerful microscopes were minimal. And there the microscope cavalry charged in to the project. By the end of my experiments, I had bombarded many species of Burgess animals with a barrage of laser and electron beams, and had imaged the specimens at extremely high magnifications – so high that even single molecules could be observed.

The techniques I used were all harmless to the fossils, which in some cases included original organic material, but there was one further test I wanted to carry out which would have altered the fossils permanently. The scanning electron microscope exacts a thin coat of metal to be applied to any animal surface under observation – a coat that cannot be removed practically. So rather than harming the invaluable fossils, casts were made. Plaster of Paris can be used to make casts of dinosaur footprints, but the Burgess Shale fossils under investigation were small and the diffraction gratings are microscopic. The particles in plaster of Paris are simply too large to fill the grooves of a diffraction grating and produce a detailed cast. But I had learnt of a new technique using acetate, and this enabled fine, elaborate casts to be made. When dry, the casts rather than the fossils were gold-coated and could be examined in a scanning electron microscope.

After the last microscopic tests had been completed, the potentially amazing became a reality. The reactions of several electron microscopy technicians indicated that the results were both positive and special. On the broken surfaces of three species – the bristle worms *Wiwaxia* and *Canadia*, and the arthropod *Marrella* – were remnants of diffraction gratings. Only traces of gratings had been preserved, rather like the few squares that remain in many Roman mosaics, but where they did occur on a single body part, they were always exactly the same size and shape, and were orientated in the same direction. The results were consistent. But the fragmentation had extinguished iridescence in the actual fossils. The fossils were decidedly grey. The mood in my lab, on the other hand, was more colourful.

Figure 6.1 Micrographs of the Burgess bristle worm *Canadia* at increasing magnification – from x10 to x1,500. The top picture shows the front half of the animal, the middle pictures show details of bristles. The bottom picture shows the surface of a bristle as removed from the rock matrix, revealing the remnants of a diffraction grating with a ridge spacing of 0.9 microns.

Did this really mean that *Wiwaxia*, *Canadia* and *Marrella* would have appeared highly coloured when they lived 515 million years ago? This still seemed unbelievable. To make doubly sure, the original surfaces of *Canadia* and *Marrella* were reconstructed in their entirety, based on the remnants that had preserved. This was achieved by carefully positioning two laser beams so they met and interfered at the surface of a light-sensitive material and etched out the precise sinusoidal contours of the remnant gratings over the entire material (the model was examined further to confirm this). The reconstructed surfaces were taken out of the dark laboratory and placed in seawater under sunlight, and . . . the colours of three Burgess Shale species shone as spectacularly as they had 515 million years ago. That was the most memorable moment of all. For the first time, the original colour of a Cambrian animal had been uncovered. An almost unimaginable piece of Cambrian history had been revealed.

When a surface has the physical properties – the size and shape – of a diffraction grating, it *will* cause iridescence in the presence of sunlight. And sunlight would have existed in the environment of the Burgess animals – at least the blue, green and yellow part of sunlight. I applied some simple optical equations to the reconstructions of *Wiwaxia*, *Canadia* and *Marrella* and calculated the directions in which they would have reflected different colours. Because the parts with diffraction gratings were positioned in a variety of orientations, from any direction *Wiwaxia*, for instance, would have shimmered with all the colours remaining in sunlight. And those colours would have appeared relatively bright like the spectrum of a compact disc. They would have been visible even under the dim light conditions of deeper waters or during dawn and dusk – when pigments become invisible. Interestingly, I photographed the model of *Wiwaxia*'s spine gratings under ultraviolet light only. Here I used the methods employed previously on the Atlas moth, as described in Chapter 3. Humans are blind to ultraviolet light, so I could see nothing through a camera with an ultraviolet-only filter. But when the ultraviolet-sensitive film was developed, very bright patterns emerged where human-visible colours were absent. The camera could 'see' ultraviolet, and I was looking at the camera's view. So if *Wiwaxia* had lived where the ultraviolet part of sunlight existed, such as in shallow depths, it would have shone brightly in ultraviolet along with the human rainbow. Unfortunately, we will probably never know the complete spectrum that illuminated the Burgess animals.

That relatives of *Canadia* and *Wiwaxia* today also have diffraction gratings is a nice test of the Cambrian finds. The spines and hairs of many living bristle worms, particularly those most closely related to *Canadia* and *Wiwaxia*, are highly iridescent. They have similar diffraction gratings and they produce colours comparable with those of the reconstructed surfaces of their Cambrian relatives. This makes the colour reconstructions of *Canadia* and *Wiwaxia* seem quite reasonable, and removes them from the realms of science fiction.

The Burgess colours quickly made the news. New scenes of life in the Cambrian were computer-generated by a number of magazine artists, but these scenes were different from those we had become used to. These were in colour, and now the colours were accurate. The Cambrian was seen as never before.

Full-colour models of Burgess creatures were also constructed in natural history museums. That ultra-impressive walk-through Cambrian reef at the Royal Tyrrell Museum also features an iridescent *Wiwaxia*, a couple of feet long of course. The addition of colour really does help to bring ancient animals to life, and now *Wiwaxia* is almost alive.

This Burgess project had certainly revealed some interesting results, but what did they mean? A standard physics textbook, Born and Wolf's *Principles of Optics*, affirms that diffraction gratings were conceived in 1819, when Joseph von Fraunhofer wound fine copper wire around a metal screw. Others credit the diffraction grating to the US astronomer David Rittenhouse, after his experiments of 1785. Now the date for the first diffraction grating has been pushed back a little further – some 515 million years. But on the serious side, some intriguing biological questions surfaced following the find of the Cambrian gratings. Why were these Burgess animals reflecting colour in the Cambrian? Was there a wide-ranging consequence to all of this? It was at this point that studies of animal colour and the Cambrian explosion first began to cross paths. It was not any old fossil that had been reconstructed accurately in colour, but one that existed relatively close to evolution's grand event.

These questions changed the course of my research and lie at the origin of this book. The book itself holds the answers. Although the finding of Cambrian colours adds nothing directly to the Cambrian enigma, it does provide a cryptic clue. And this was the first clue that I uncovered, which ultimately led to the writing of this book.

Up to this point the chapters in the book have contained the thoughts that go through one's mind, in the order they happen, while contemplating the questions that followed the Cambrian colour discovery. But there are further thoughts to be introduced, involving subjects that make up the final pieces of the Cambrian jigsaw puzzle. These will be covered in the next two chapters; the first of these subjects may indeed seem overdue.

So much discussion of colour warrants consideration of its counterpart. There is a reason for the variety and sophistication of the colour we see today; 'see' is the operative word. One particular organ exists that conceives both the observer and the observed – the eye.

7

The Making of a Sense

> To suppose that the eye, with all its inimitable contrivances . . . could have been formed by natural selection, seems, I freely confess, absurd in the highest degree
>
> CHARLES DARWIN, *On the Origin of Species* (first edition, 1859)

The preceding chapters have explained and emphasised the importance of light as a powerful stimulus to animal behaviour in the past and present, and revealed it as a driving force of evolution and a promoter of great biodiversity. This chapter is devoted to the eye and the reason for this influence of light on animals and their evolution – the sense of *vision*.

Eyes are the detectors that convert the light waves travelling through the atmosphere into visual images. These light waves enter the Earth's atmosphere from the sun, and bounce and reflect off objects that exist all around us. They are the same light waves that change when they strike an animal to relay information about its identity and whereabouts within the environment. Eyes pick up all this information. Eyes and only eyes conceive the sense known as vision. Electromagnetic radiation of different wavelengths exists in the environment; colour exists only in the mind.

In Chapter 4 I questioned whether the Precambrian environment was similar to that found in caves today. By the end of this chapter we will be able to link light, eyes and vision, and understand that such a question is not well founded. We have established that the Earth is said to be 4,600 million years old, as is the sun. So sunlight would, to some degree, have struck the Earth's surface well into the Precambrian – but it would not have entered caves. Not now, not then. Moving from here to the next question I will pose, we will approach the final solution to the Cambrian enigma. Two further clues remain to be found in Chapters 7 and 8, and these will provide the final pieces of the Cambrian puzzle. For the moment, however, we can look for a more immediate clue in the question: '*When* did eyes invent vision?'

Before attempting to answer this specific question, a tour of the wide range of eyes found today is necessary if only to interpret fossil eyes. Darwin referred to the eye as an 'organ of extreme perfection and complication'. The word *eye* implies an organ capable of producing visual images in order to distinguish objects using light. Extreme perfection and complication are obligatory characters of the more efficient eyes, and so the reference in Chapter 4 to the eye being a very expensive piece of equipment is really quite valid. But the eye itself is only Act One in the complete performance of seeing. Act Two involves transmitting visual information, in the manner of electrical cables, from the eye to the brain. In Act Three an image is formed in the brain. Vision employs the eye *and* brain of the beholder.

The central aim of this chapter is to trace the introduction of the eye to Earth. Since only the eye is preserved in fossils, and not information relating to Acts Two and Three of the visual performance, this chapter will centre on the architecture of the eye itself – the main hardware. We will assume that an eye with good optical apparatus is linked to a brain where a good image is formed, and a poorly designed eye to a brain producing poor images. In other words, the complexity in the hardware is mirrored in the software. Only the box jellyfish can throw a spanner in the works of this theory, but the box jellyfish is destined to emerge as an oddball anyway.

Vision – the formation of an image or picture from light waves – is the most sophisticated form of detecting light, but it is not the only one. The less sophisticated, or elementary, forms are relevant to Precambrian life, and so to the theme of this book. The elementary form of detecting light will be called 'light perception', and the receptors that perform this task 'light perceivers'. The question of interest in the first part of this chapter is 'To see or not to see?' Throughout the remainder of this book, it is vital that these two possibilities and their associated organs are kept very separate.

Not to see

Light perception in bacteria, animals and plants ultimately involves organic molecules that undergo a simple reaction when hit by a package of light called a photon. Light perception takes place in many single-celled animals, such as amoebae and *Euglena*, where the fluid within the cell is sensitive to light. These animals use light to orientate themselves – to distinguish up from down.

In multicelled animals, independent light-sensitive cells or organs of various complexities perform the task of light perception. The most elementary forms of light-perceptive *organs* are called ocelli. These are small cups containing a light-sensitive surface backed by dark pigment. Sometimes they are capped by a rudimentary lens. The simplest multicelled animals with these structures are the jellyfish.

The marginal sense organs of jellyfish in some cases include ocelli, in addition to gravity, touch, chemical, pressure and temperature receptors. Indeed, ocelli are generally the most poorly developed sense receptors in jellyfish, with lenses lacking from most groups. The pigmented patches of most jellyfish are not known to detect light, and may have evolved rather as a light barrier – to absorb light and so shield the underlying sensory cells that detect other stimuli. But in some jellyfish, where a lens covers the cup-shaped light-sensitive surface, the ability to respond to light on or light off has been established.

Similar cup-shaped ocelli occur in members of many other animal phyla such as flatworms, ribbon worms, bristle worms, arrow worms, molluscs and sea squirts. An advantage of a cup-shaped light perceiver over a flat one lies in its curved surface. A beam of sunlight illuminates a curved surface, such as a hemisphere, in one region only. A flat surface, on the other hand, would be completely lit. So a curved surface can perceive the direction of the light source. Some maggots – the larvae of flies – possess flat light perceivers but still manage to find a light source by swinging their heads from side to side. This mechanism, not surprisingly, is uncommon.

The elementary light detectors discussed so far cannot be called eyes because they don't form images. Eyes are born when the light detection cells get serious and form a 'retina', a thin plate of nerve cells lining the

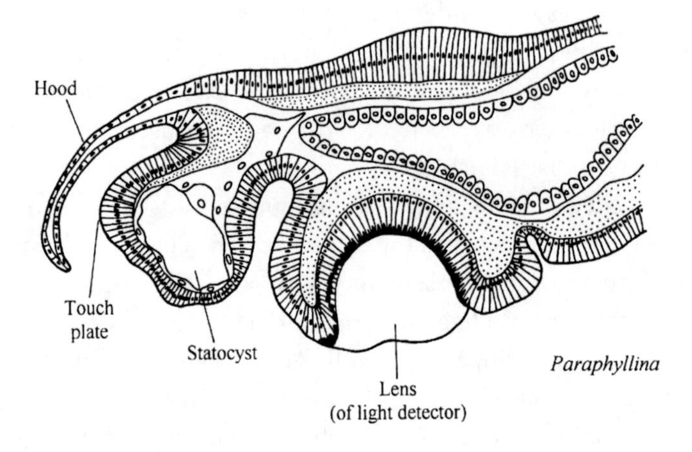

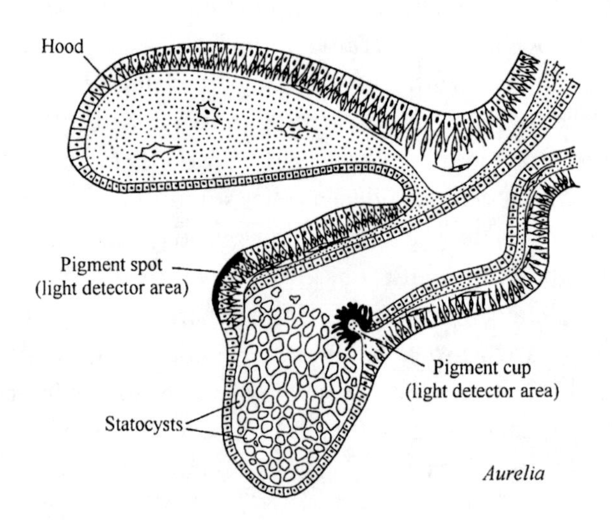

Figure 7.1 Marginal sense organs of the jellyfish *Paraphyllina intermedia* and *Aurelia aurita*, showing different levels of complexity (particularly in their light detectors).

inside of the eye. The retina will detect with accuracy whatever is projected on to it, so it is important that an image is first focused sharply on to the retina by some additional apparatus. A camera loaded with highly sensitive film would be useless without a lens. When all these conditions are satisfied, we have an eye – we have reached the stage of being able 'to see'. And the size of the step taken to get here cannot be overemphasised.

Based on the number of entrances for light, eyes can be divided into two types – simple and compound.

To see

'Simple' eyes

Simple eyes are so called because light is received through a single entrance – the simplest design solution for an eye . . . in theory. Molluscs may exhibit a wide variety of light perceivers, or 'eyespots', but they also boast the broadest range of eyes. And these are all simple eyes. But despite their inept title, simple eyes do produce visual images, and ironically their hardware is often quite intricate. There are three forms of simple eyes known in animals, and all can be found in molluscs.

Nautilus, the subject of a palaeontological mystery discussed in Chapter 2, has a simple eye that is unique because an image is produced on its retina without the aid of a lens. For more than 2,000 years the Chinese have known that an inverted image is produced on the inside wall of a dark chamber if light enters only through a small hole in the opposite wall. Leonardo da Vinci revived this principle with his 'camera obscura'. But the Chinese had, unknowingly, revived it too – the principle was practised by nautilus long before.

The image-forming structure in the 'pinhole eye' of nautilus is a small pupil, or 'pinhole', formed in its dark iris. Light is not focused, but is received only through the pinhole, providing at least some degree of control. To gain accurate directional information, the retinal mosaic is remarkably fine so that light coming from a single point will illuminate several receptor cells. But serious disadvantages are inherent with this type of eye, which accounts for its rarity. A bright image requires a large pupil, whereas a sharp image requires a small pupil. The nautilus' solution – a large range of pupil sizes or pinholes – unfortunately results in blurred images.

In his book *Optiks*, published in 1704, Sir Isaac Newton revealed his plans for a telescope without a lens but with a curved, concave mirror instead. This mirror would focus light towards a focal point, in the same way that a modern satellite dish focuses radiation towards its receiver. At the focal point was positioned a small, flat mirror, angled to redirect the focused light out through a gap in the side of the telescope – the eyepiece. This 'Newtonian' telescope works well – it is popular today.

Curved mirrors can also be successful substitutes for lenses in eyes. The scallop has many eyes just inside the edge of its shell. These eyes appear silver, like tiny mirrors – and indeed they do contain mirrors. Within each eye, a hemispherical concave mirror similar to the reflector in a car headlight lies *behind* the image-forming retina. Light passes almost unfocused *through* the transparent retina before it is reflected back, this time focused by the mirror. The light is focused precisely at the position of the retina. And now the retina absorbs the light rays, and an image is grabbed. The mirror is achieved by the same mechanism found in the skin of the Mexican cave fish – stacks of thin layers of various thicknesses. The mirror eye is an improvement on the pinhole eye because it can focus light. But with light first passing through the retina unfocused there is potential for *this* to be detected, and so the performance of the eye is limited. For this reason, the mirror eye is confined mainly to the scallop and a few related clams.

The third type of simple eye in molluscs is found in a snail. The snail has an eye separated from the skin and containing a large spherical lens. This eye is known as the camera-type. It works in the same way as a camera in that a single lens focuses light on to a film, or retina, with an adjustable iris included to alter the quantity of light passing through its 'pupil'. The general design is quite simple, but it is ideal for seeing images, and the variety of camera-type eyes to be found in other animal phyla testifies to its success.

The most efficient eyes of bristle worms belong to a group known as

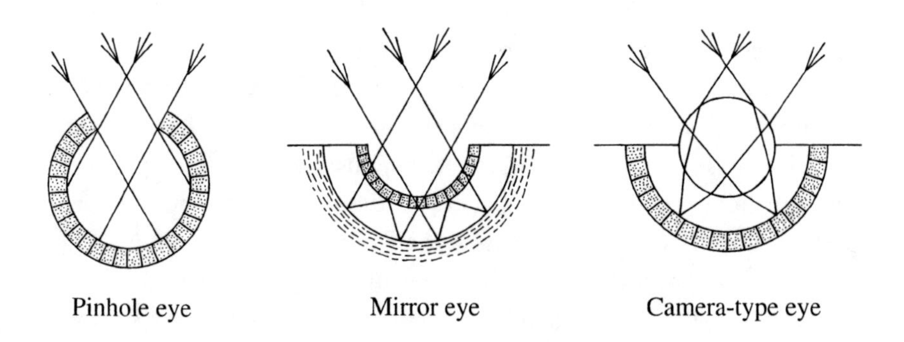

Figure 7.2 The three types of simple eye – pinhole, mirror and camera-type – and their effect on light rays. Light receptors (retinas) are shaded. The mirror eye has an underlying mirror (dashed region) and the camera-type eye has a lens, both of which focus light to form clear images.

alciopids. One member of this group lives on the surface of the sea and possesses camera-type eyes complete with a paired retina with refined layering, two distinct layers of 'humoral' material which fills the eyeball, and a well-developed spherical lens and 'cornea' – the outer covering of the eye. The retina contains about 10,000 light detection cells, and is positioned at the focal plane of the lens – the position where the lens focuses an image.

The camera-type eye is the standard hardware for vision in vertebrates, both on land and underwater. Humans are one beneficiary, but in addition to bristle worms and molluscs it has also emerged in spiders and crustaceans within the arthropod phylum, velvet worms within a phylum all of their own, and in box jellyfish within the cnidarian phylum. The precise design of the camera-type eye is determined by how the lens is formed – it can be formed either inside the eye, or outside, where it is actually part of the skin or exoskeleton, and is technically the cornea.

Focusing is all about bending light rays from different parts of the environment towards a common point. There are two factors which affect the bending of light rays – the differences in materials either side of a boundary, and the angle of that boundary relative to a light ray (think of a prism). Adaptations to vision on land are different from those underwater because, as we learnt in Chapter 3, light behaves differently in air compared to water – there is a material difference. Light does behave similarly in water and in the cornea, so it barely recognises a boundary as it enters the eyes of aquatic species. In this case, the lens within the eye must be responsible for most of the focusing. But light recognises a considerable difference between the cornea and air, and it is bent as it crosses their boundary at an angle. So the cornea of land animals acts as a powerful lens in its own right.

James Clerk Maxwell tackled the subject of underwater focusing in the nineteenth century while contemplating his breakfast herring. Following a spontaneous dissection, Maxwell noticed his herring had a spherical lens. This is typical in fish – it bends light rays more than a thinner, oval lens because its surfaces are more steeply curved and present steeper tangents to light. But there is a problem with a spherical lens – spherical aberration. This is the reason why we do not use spherical lenses in cameras, and instead choose a series of 'oval' lenses. Spherical aberration occurs when light striking the periphery of a lens is focused at a different plane to light striking the central axis of the lens – the peripheries bend light too much. So to focus both sets of rays simultaneously, the retina must be in two places at the same time. This is impossible. But fishes *can* see by focusing very sharp images – the question is 'How?' There is only one solution here, and Maxwell worked it out.

If the curves, or rather tangents, are made less steep by flattening the lens, the focal point moves too far from the lens – a huge eyeball would be needed to house the retina. So if the angles can't be changed there is only one other option for solving spherical aberration – change the materials. And indeed Maxwell suggested that the material of the fish lens is not uniform but is graded from the centre outwards.

Today we know from precise measurements that the periphery of the fish lens has optical properties similar to those of water and causes light to bend only slightly. This compensates for the comparatively glancing angle with which light strikes the edge of the lens, which alone causes considerable bending of the light path. Near the central axis, this angle is not glancing but nearer to 90°. So to keep central light rays synchronised with those from the periphery of the lens, the lens material in the centre is optically very different from water and causes light to bend more, but also to slow down more. This effect on the speed of light is important since the path through the centre of the lens is now the shortest. To sum up the effect of this lens, *all* the light rays striking the eye at one instant will be focused on to the same point on the retina at the same time. Clever! Now a sharp image is formed . . . and formed equally well in all directions.

Not surprisingly, the graded lens has been an evolutionary success in the water – it is also found in marine mammals, tadpoles, and some molluscs such as octopuses, squids, cuttlefishes, winkles, conches and pond snails. Michael Land, an authority on the eyes of living animals

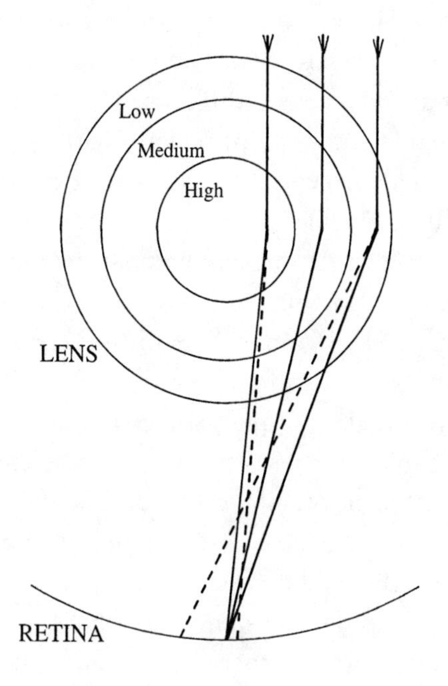

Figure 7.3 Focusing of light rays (solid lines) by a graded lens (only three grades of material are shown). The dashed lines represent the paths induced by a standard lens of uniform material – the steeper angles of contact cause light to be bent more at the periphery. The core material of the graded lens, however, causes light to bend more than the material of the periphery layer, and so counteracts this angular discrepancy.

based at Sussex University in England, calculated that if nautilus had a camera-type eye of the same size as its pinhole eye, it would be four hundred times more sensitive and have one hundred times better resolution. Sensitivity refers to the ability of an eye to get enough light to the receptor cells, whereas resolution is the precision with which light rays from different directions are kept separate (to prevent blurring).

In eyes of vertebrates on land, between 20 and 67 per cent of the focusing power is supplied by the cornea. So the lens can be designed specifically to correct blurring and to make accommodation for different distances – nearby objects would otherwise be imaged further away from the lens than distant ones. Mammals, birds and most reptiles meet these goals by changing the shape of their lens or cornea, thereby adjusting the position of the focal plane. They use tiny muscles to pull and stretch the lens. Alternatively, fish, frogs and snakes move their lenses backwards and forwards. Lens movement can provide adjustments for water or air in some amphibious animals.

Although most spiders are equipped only with ocelli, jumping spiders and wolf spiders are exceptions. They have camera-type eyes with a thick cornea, which provides all the focusing power. The principal eyes of the jumping spider are unusual because, like birds of prey, they have a large pit in each retina that acts as a negative lens – it inverts and magnifies the image. This is comparable to the rear element of a telephoto lens in a camera.

The principal eyes of the jumping spider are also unusual because their retinas are narrow vertical strips, giving a restricted visual field of only a few degrees in the horizontal direction but about 20° in the vertical. These retinas move sideways to compensate for their thinness, and so greatly extend the field of view in the manner of a photocopier scanning a picture. A similar mechanism is found in some snails and crustaceans. In some other crustaceans, such as the iridescent *Sapphirina* with the appearance of a swimming opal, there is a more extreme development. Here the retina is merely a dot with only a few light detection cells, but constant movement in all directions puts these cells to continual use.

The iris controls the pupil size and, consequently, the amount of light entering the eye, in the manner of a camera diaphragm. But a further mechanism for brightness control may also exist in camera-type eyes, involving a reflector lying behind the retina. Like the scallop eye, the eyes of some vertebrates also contain a mirror behind their retinas, again employing the silver fish–mirror mechanism. But here the mirror functions so as not to focus the light – light is prefocused by a lens. In this case, the mirror provides an adaptation to the night. In the dark-adapted state, the reflector returns to the retina light rays that initially passed between the retinal cells without detection. So the most is made of the light available – anything not detected first time will have a second chance. The reflector in the eyes of cats and crocodiles reflects the beams of headlights and torches, often appearing obvious as 'eye-shine' at night. This represents the light missed both first *and* second time by the retina. When light levels are very low, *all* the rays striking the eye become invaluable to vision – the visual frontier between sight and blindness is approached. But when light levels are high, the reflector is redundant and becomes covered by a dark absorbing pigment. This mechanism is actually common in many nocturnal animals with camera-type eyes.

Briefly stepping back over the visual threshold, into 'not to see' territory, some light perceivers actually contain a retina and a lens. Included among these are the receptors of scorpions, many web-building spiders and most snails. But their small size is important. This receptor type is not an 'eye' because it cannot focus an image on its retina – the retina is simply too close to the lens. The likely function of this receptor is to measure the average brightness or colour over large angles. As mentioned already, we are not so concerned with these light perceivers in this book, but they do help to illustrate just how much information can be acquired just from the architecture and size of a structure containing a lens and retina. The reason for writing about today's eyes in this chapter is merely to provide a palaeontological tool. Where internal views of a simple eye-like structure are possible, we can calculate whether visual images were formed – whether it really was a simple *eye*. And the *size* information will become even more important when we consider fossils in relation to the Cambrian enigma. But fossil eyes that reveal internal architecture are extremely rare. So the best studied fossil eyes belong to the second type of eye found in animals, one where more information on sight can be deduced from just an outer surface. At least half the animals on Earth today are equipped with compound eyes.

Compound eyes

Figure 7.4 Scanning electron micrograph of the head of a fly, showing compound eyes.

By way of introduction to compound eyes, I will return briefly to the somewhat unrefined ocelli. There is one group of bristle worms with ocelli but these are different from those of other animals. The difference lies in their arrangement. Lying on thick, feather-like filaments sprouting from the head, the ocelli of these worms are grouped together.

Each ocellus has a sac-like region formed as an outgrowth of a sensory hair. This region lies within an infold in the skin of the animal and acts as a lens. Behind this lies a well-developed region of light-sensitive chemicals – the 'retina'. And within a group of ocelli, light-absorbing pigment cells intermingle to prevent the same light rays affecting more than one ocellus. But the information collected by each ocellus is later combined strategically and so elaborate composite organs are formed. An organ of this type is known as a compound eye (although these particular eyes fall a little short of the visual mark).

In contrast to the simple eye, the compound eye has multiple openings for light to enter – hence its name – and so always consists of numerous individual units, or ocelli, called 'facets'. Other than minor appearances in the bristle worms and ark clams, the compound eye is a character of the arthropods. More precisely, compound eyes today occur in crustaceans, insects and horseshoe 'crabs' (which are actually more closely related to scorpions than true crabs). Compound eyes have evolved into sophisticated organs of sight, up to a third of the total body size in some seed-shrimps, and form images in different ways.

The law of compound eyes was laid in 1891 in a monograph by the biologist Sigmund Exner, which became a landmark to both biologists and optical theorists. Exner broke all the rules of his day, where simple eye concepts were being applied to compound eyes. Instead Exner considered the focusing elements of compound eyes as 'lens cylinders'. A conventional lens relies on the bending of light as it crosses a curved surface to focus rays. A lens cylinder, on the other hand, gently persuades light to change direction throughout the cylinder's length. It is literally a cylinder, but one filled with graded material – graded in terms of its effect on light, just like the fish lens contemplated by Maxwell. The lens cylinder is most dense, and so causes light to travel slowest, along its central axis, its ability to bend light fading towards the edges. The overall effect of a lens cylinder is to provide the image-forming properties of a traditional lens. But there are alternatives to lens cylinders in some compound eyes.

The compound eyes of many insects and crustaceans have a similar superficial appearance, but their focusing elements and mechanisms of image formation are very different. We can divide compound eyes into two basic types – apposition and superposition. The facets of apposition eyes are optically isolated from each other, so they each sample a different section of the environment. The tiny images formed within each facet are pieced together in the manner of a jigsaw puzzle to produce the complete picture. The facets of superposition eyes, on the other hand, cooperate optically so that they superimpose their light to form a single image at a common point on the retina. Dividing compound eyes further, there are variations of both apposition and superposition compound eyes in terms of focusing and image formation.

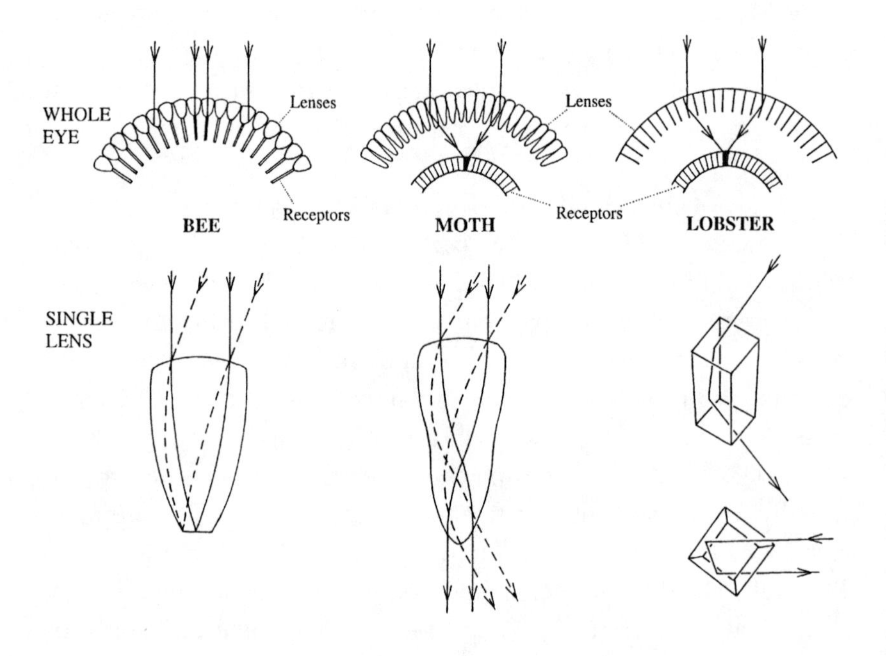

Figure 7.5 Focusing mechanisms in the compound eyes of a) bees (apposition-type eye); b) moths and c) lobsters (superposition-type eyes). Graded material in b) and mirrors in c) (shown from the side and from above) achieve focusing. (Modified from Land, 1981.)

Superposition eyes contain a large number of facets – up to several hundred. A broad zone of clear material separates the lenses from the retina below – the facets are not continuous tubes, so light can cross from the lens of one facet to the receptor of another. Exner discovered this after considering the cornea of the glow-worm (actually a beetle) as a single structure rather than as individual units. The cornea forms the lenses of the glow-worm, and Exner simply cleared out the interior of a glow-worm eye so that he could examine the complete array of lenses. Instead of a series of inverted images corresponding with each lens, he found just one upright image. So all the lenses must focus on to the same point on the retina.

Some crustacean eyes are without lens cylinders, and it was not until 1975 that an alternative focusing mechanism was discovered. While examining the eyes of a crayfish and a deep-sea shrimp respectively, Michael Land and the German biologist Klaus Vogt independently found a superposition eye in which each facet was lined with mirrors. The mirrors were again similar to those found in fish skin, and formed mirror boxes, square in section. Exner got as far as illustrating the shape of these boxes, but overlooked the silver inner surfaces. It is now clear that if the boxes are considered with mirrored sides, the light rays will change direction as they are reflected and will all meet at the same point – on the retina where they form an image. In other words, mirror boxes perform the task of focusing.

In 1988 a third form of superposition eye was discovered in many crabs by another contemporary expert on eyes – Dan-Eric Nilsson of Lund University in Sweden. The optics in this case are complex, and involve an intricate combination of ordinary lenses, cylindrical lenses, parabolic mirrors and light guides. The imaging mechanism is equally elaborate, with three separate systems in operation. Image formation can be predicted from the hardware alone, suggesting that fossil eyes could be informative after all.

Compound eyes have no iris to control light levels, but they do have an alternative solution – they use dark pigment to remove some light where needed. This is similar to the way cats and crocodiles use pigment, but the pigment is in a different part of the eye. When superposition eyes are exposed to high light levels, the dark pigment moves between the lenses and retina to absorb a proportion of the light rays. When light levels become extreme, sometimes the facets become optically separated by dense collars or tubes of absorbing pigment so that the eye effectively acts as an apposition eye.

All in all, the architecture or optics of modern eyes are well understood and can teach us much about the way their hosts see. And architecture can be preserved in the remains of extinct animals. We are now well enough informed to be able to browse the library on fossil eyes.

Ancestral eyes

The post-Cambrian view

Conodonts are animals named after the Greek word meaning 'cone teeth'. This is because for some time they were known only from jaw-like structures and bone fragments. Conodonts evolved in the Cambrian and became extinct 220 million years ago. They have been used extensively for comparing and assessing the age of rock sequences, but until the early 1980s we had no idea what the conodont animal looked like. Then complete fossil conodonts, around 340 million years old, were found in the Granton Shrimp Beds near Edinburgh, Scotland. These fossils revealed animals of eel-like appearance, with tails containing supporting fin rays . . . and heads with large, camera-type eyes.

The smaller species of conodonts have eyes which are larger in relation to their body length than are the eyes of the larger species of conodonts in relation to theirs. This is consistent with a rule of the relative growth of the eye dating back to 1762, which holds that smaller animals have larger eyes in relation to their body size than do larger animals. From work on living vertebrates, eye size is known to influence visual sharpness. And the fact that conodonts possessed reasonably large eyes carries important information on early vertebrate evolution.

Theories that conodonts are larval stages of 'agnathans', the primitive jawless fishes that include today's lampreys and hagfishes, have not been well received. Agnathans are the first representatives of the true vertebrates, within the chordate phylum, which evolved around 485 million years ago, just after the Cambrian. But the relatively small eyes and large bodies of some conodonts are evidence against their interpretation as juvenile agnathans. So conodonts are more generally believed to be chordates close to, but not part of, the line *leading* to true vertebrates. The eyes tipped the balance of opinion in this direction.

Among the living agnathans, only hagfishes have 'eyes' that are smaller than those of conodonts. But the 'eyes' of hagfishes, those primitive fishes caught in the deeper-water SEAS traps, are rather light perceivers – they do not yield visual images. They have probably degenerated as an adaptation to dark environments and burrowing. On the other hand, the eyes of lampreys are well developed and generally larger than those of conodonts. But there is one group of lampreys – the smallest brook lampreys – which offers clues to the vision of conodonts.

Small brook lampreys have eyes that are about a millimetre and a half in diameter, equal in size to the eyes of the conodont *Clydagnathus*. There is evidence to suggest that similarity in the size of camera-type eyes reflects a similarity in cell and nervous complexity – the information processing system. The finding of eye muscles in another, well-preserved conodont, *Promissum*, supports this line of thinking – the muscles of similar-sized eyes are also equivalent.

The conclusions drawn from a comparison with small brook lampreys are that conodonts had pattern vision and, as will become relevant later in this book, an active predatory lifestyle. Nevertheless, to say that smaller conodonts had relatively larger eyes does not imply that vision played a greater role in their behaviour. Instead they possessed visual organs near the minimum size limit for 'eyes' – these organs could not have produced visual images if they were smaller. *Thorius*, a miniaturised salamander which is the smallest land vertebrate living today, has camera-type eyes that are little over a millimetre in diameter, and this is believed to be the lowest size limit that will provide precise vision.

At the larger practical limit of eye size was *Ophthalmosaurus*, a dolphin-shaped reptile between 3 and 4 metres in length. While dinosaurs were evolving big bodies on land, *Ophthalmosaurus* was setting a record for camera-type eyes in the sea. This animal really did possess eyes the size of soccer balls, and they were used to see at depths of 500 metres and more. It is thought that this reptile dived deep to avoid predators or to catch deep-dwelling prey. Unfortunately *Ophthalmosaurus* suffered from 'the bends', a condition familiar to deep-sea divers who approach the surface too rapidly. Rapid ascent causes nitrogen gas dissolved in the blood to decompress, forming bubbles that can block blood vessels and kill tissue. The bends leave visible depressions in the joints of bones, and those depressions are evident in the fossils of *Ophthalmosaurus*. The bends, and its effects, occurred much less commonly with its smaller-eyed ancestors, but the deep-diving, big-eyed version was stopped in its evolutionary tracks – and so the *Ophthalmosaurus* died out.

Remaining within the ancient vertebrates, early fossil fishes commonly have dark stains in the region of the head. But what do these stains represent? The eyes of one particular agnathan specimen can provide an answer. This fossil fish has, in the exact position of the stains found in other agnathans, an eyeball fully preserved in a subspherical hardened structure with a slit directed sideways. Alex Richie, an expert on primitive fishes at the Australian Museum, has interpreted the stains as the remains of pliable, presumably cartilaginous, structures surrounding the actual eyeball. The earliest of these camera-type eyes are known from a 430-million-year-old specimen of *Jamoytius kerwoodi*. Richie believes that although not specifically found in the eyes of this fossil specimen, there is no doubt that a lens of some form was present, based on comparisons with these camera-type eyes found in later, related species.

So eyes have been found in fossils up to 430 million years old, and their vision has been extrapolated via comparisons with the eyes of today. But can we wind the clock back further and take a look into even older eyes?

Back to the Burgess Shale

After scrambling up a slippery section of the Canadian mountainside from the camp of Des Collins's field team to the Burgess Shale quarries,

I reached a ledge that had been excavated during both earlier and current fossil expeditions. At the back of the ledge was the exposed face of the quarry, where the layers of sediment were clearly defined through their various colours. On the ledge itself was a wooden table, which supported the fossils unearthed during the current excavations.

Standing in the viewing area of the Burgess quarry, on the edge of Emerald Lake below, all that could be seen of the site above us was an indiscernible blue object. Along with the many tourists using the telescope provided, I wondered what it could be. It was, as I discovered on reaching the quarry, just an old plastic sheet, but one with an important purpose – to protect the latest fossil treasures laid out on the table from the harsh climate. These fossils were awaiting quality control, to ascertain their future in the display cases of museums all over the world. And there really were some treasures. The fossils whose photographs I had seen in coffee-table books and on the projector screens at a number of famous lectures were there in front of my very eyes. And I was one of the first people to see these specimens – they were fresh from the rock. But they were very well preserved and defined – and I could identify them all.

I picked up the largest piece of thin, flat shale on the table. It had the dimensions of a large roof slate, and its smooth surface bore a detail of the most fearsome member of the Burgess community – *Anomalocaris*. The body was big, nearly half a metre long and broad with it. Emerging from the head, the grasping forelimbs, once thought to be shrimp-like animals in their own right, were obvious. And I had already identified the front end of the body thanks to another give-away clue – the large pair of eyes that were equally obvious.

The eyes of *Anomalocaris* appeared as two buttons jutting out from the sides of the head. Their smooth, rounded outlines were obvious, although that was all to be seen with the naked eye. But their position on the sides of the head suggested these were eyes and nothing else. In Chapter 1 we learnt that no new animal phyla have evolved on Earth since the Cambrian explosion – the phyla we see today are those that existed in the Cambrian (with few possible exceptions). There is a law also that animals today live and function as did their ancestors in the Cambrian. There have been no magical periods in history since the

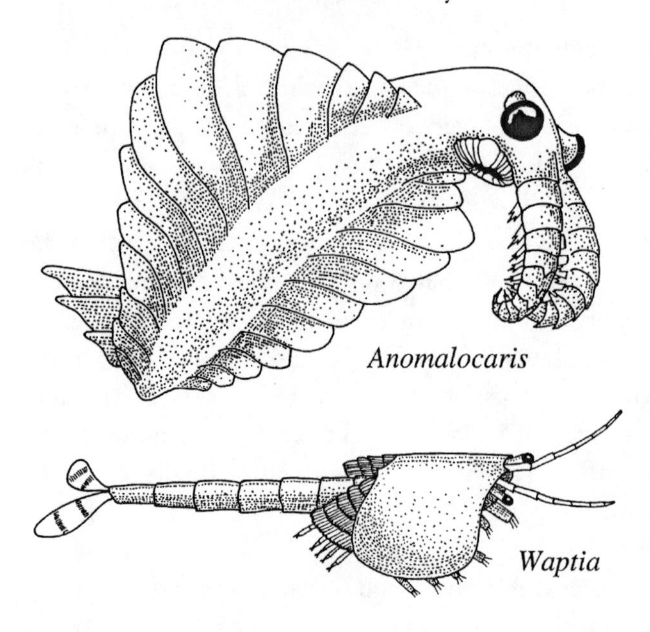

Figure 7.6 *Anomalocaris* and *Waptia* from the Burgess Shale. At around 7.5cm, *Waptia* is several times smaller than *Anomalocaris*.

Cambrian explosion where things happened differently. Today it is clear that the button-like structures protruding slightly from the head of *Anomalocaris* could be only one thing – eyes.

Back in the laboratory at the Smithsonian Institution, I examined the well-preserved analogues of another member of the Burgess community – *Waptia*. *Waptia* was a shrimp-like animal, a member of the arthropod phylum and possibly a crustacean. It was also about the size of an average shrimp of today, and seemed to share the shrimp's eye characteristics. Like shrimps and crabs, *Waptia* had eyes on stalks. This means that its eyes could have moved independently of its head. They would have been specialists at looking within a narrow range of the Cambrian environment in detail. They would have seen *Anomalocaris* as it swam in front of them. But as *Anomalocaris* moved, the eyes of *Waptia* would have moved too, and followed their giant neighbour. Unlike the compound eyes of insects, which are known as sessile because they are fused with the head and so cannot move independently, stalked eyes can change their field of view without head movement.

I mentioned 'compound' eyes during this discussion of *Waptia*. Although the eyes of the *Anomalocaris* I examined revealed little additional detail under the microscope, a microscopic view of a well-preserved *Waptia* specimen told a different story. The internal architecture

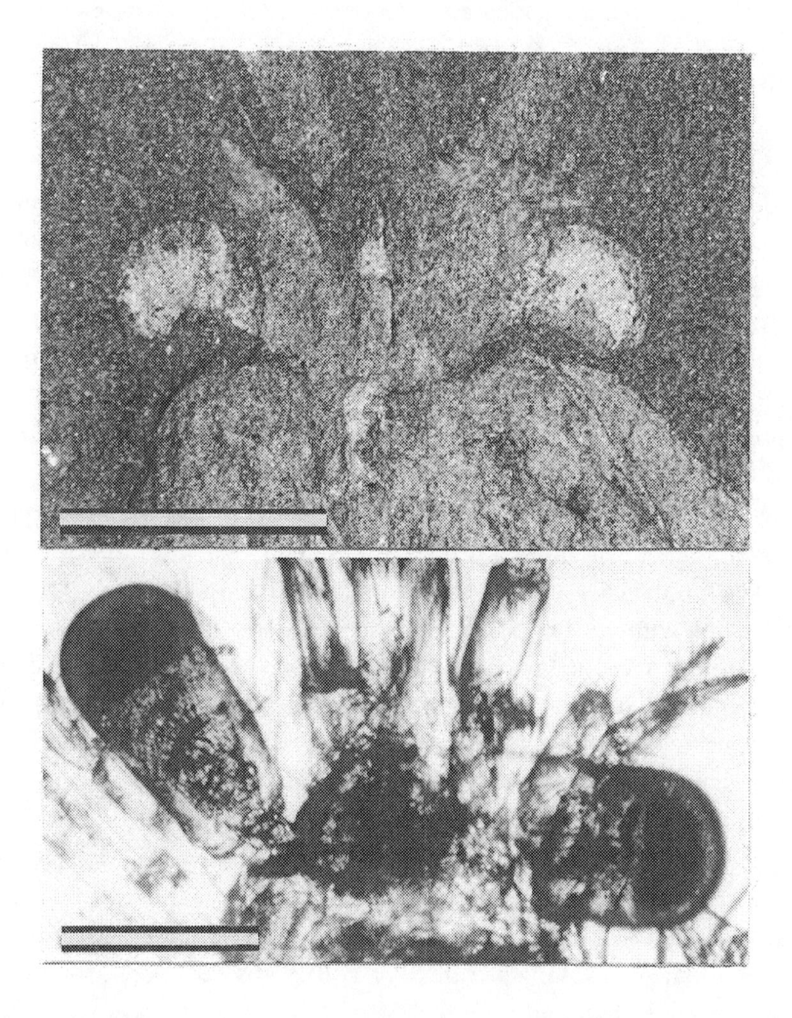

Figure 7.7 Micrographs of the heads of a living 'mysid' crustacean and *Waptia* from the Burgess Shale. Eyes show comparable internal architectures. Scale bars represent 2mm (top picture) and 0.5mm (bottom picture).

of the *Waptia* eye became evident – and it matched that of a crustacean today. The stalked apposition compound eyes of a crustacean known as a 'mysid' are producing images of animals swimming past them in the sea today. And *Waptia* would have seen similar pictures in Cambrian seas. *Waptia* had apposition compound eyes.

Looking through the collection of arthropods in the Smithsonian's Burgess Shale collection, it became obvious that *Anomalocaris* and *Waptia* are not alone. They were not the sole beneficiaries of vision: far from it.

Within the Smithsonian fossil deposit, the Burgess specimens are enclosed by a large metal cage, which provides additional security in the style of a bank vault. Doug Erwin is custodian of Charles Doolittle Walcott's collection today, which embraces quite a variety of multi-celled animal forms. Doug kindly allowed me use of his microscope, a key to the Burgess vault, and a large wooden tray.

Examining the invaluable fossils was time-consuming. They were stored in dozens of cabinets, with hundreds of drawers full of specimens. I looked into each drawer and tried to select the best-preserved representatives to fill my tray. This is difficult to do with the naked eye, and I probably missed some informative examples.

When I had made each selection, the individual fossil was placed in my tray and an official museum form was put in the empty space in the drawer. On the back of each fossil was painted a catalogue number, and this, along with my name and the name of the specimen and its original location was recorded on the form. The level of security and guardianship echoed that at the Burgess quarries themselves. The small quarries are approached by only one path – there are no back doors on the exposed mountainside. In the vicinity of the quarry, the path is policed by Des Collins's field team, whose camp is positioned just the other side of the path to the Burgess quarries. The two exits of the path, each at least three hours hike from the quarries, are patrolled by wardens from Parks Canada. And the security pays off. The world fossil trade is a considerable one. There are many private fossil collections and shops around the world. Some include complete skeletons of dinosaurs, such as *T. rex*, but none contains a single specimen from the Burgess Shale.

I examined specimens of the Burgess arthropods *Canadaspis*, *Odaraia*, *Perspicaris*, *Sanctacaris*, *Sarotrocercus*, *Sidneyia* and *Yohoia*.

All possessed eyes, varying in size with respect to body length. Again, the smaller specimens appeared to have relatively larger eyes. And all these 'eyes' really were eyes; based on comparisons with the visual organs of living species, they would have formed images in the Cambrian. In many other Burgess arthropods, the presence or absence of eyes could not be resolved with accuracy due to imperfect preservation or unfavourable orientations within the rock. Maybe I had chosen the wrong specimens to examine. For instance, I failed to detect the eyes of perhaps the commonest Burgess arthropod, *Marrella*. Recently Des Collins and his Spanish colleague Diego Garcia-Bellido identified eyes on *Marrella* resembling those of woodlice today. But I was certain of one thing – eyes were common in the Burgess arthropods.

There are a few Burgess animals from other phyla with eyes, but not many. Actually there may be only *Nectocaris* and the weird, five-eyed *Opabinia*. But then *Opabinia* is probably an arthropod, although *Nectocaris* appears closer to the chordates than the arthropods. More specimens of these rare species are needed in order to classify them with certainty. But eyes are either rare or absent in the non-arthropod Burgess animals.

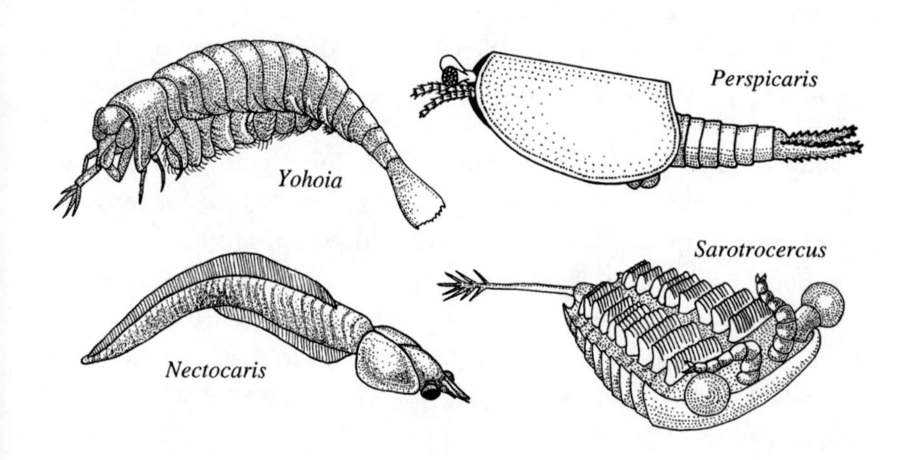

Figure 7.8 *Yohoia, Perspicaris, Nectocaris* and *Sarotrocercus* – examples of Burgess animals with eyes.

The Burgess Shale community lived in the Cambrian, but more precisely they lived 515 million years ago. The question we would most like to answer in the chapter is 'When did the first eye appear on Earth?'. Now we know that eyes were well in place on Earth some 515 million years ago, but the Cambrian explosion took place sometime between 543 and 538 million years ago. So at this point I will leave the Burgess Shale fauna and continue my search for eyes in other, older fossils (I hope) from the Cambrian period.

Other Cambrian eyes

On the subject of weird-looking Cambrian fossils, *Cambropachycope* and its relatives are bizarre arthropods that were the ancestors of the crustaceans of today. They are known from Cambrian fossils preserved in the 'Orsten' limestone of Sweden. Their preservation is actually quite exceptional, and in full 3D. The German palaeontologist Dieter Walossek is responsible for the excellent interpretations of these fossils, and as their guardian he obligingly sent a specimen of *Cambropachycope* to me for examination in electron microscopes. I was interested in this animal for one reason in particular – its eye. I use the singular here because *Cambropachycope* had a huge visual organ compared with its body size . . . but only one of them.

Cambropachycope was a small arthropod, just a few millimetres long. Its body is distinct for having a big, paddle-shaped limb on either side, so swimming may have been possible. The head of *Cambropachycope* is just as unusual. It is fused with the rest of the body, and has an obvious mouth near the fusion. But then it constricts to form both a false 'neck' and a huge bulbous projection in front of the body and mouth. The projection is basically a compound eye. Maybe it evolved following the fusion of two stalked eyes, but it certainly would have seen whatever was ahead of it with some accuracy. I drew this conclusion after studying its cornea – unfortunately that's all that remains of this eye.

Although from the Cambrian, *Cambropachycope* and the other Orsten arthropods with eyes are no older than the Burgess Shale community. But, as we learnt in Chapter 1, there is a site in China where an

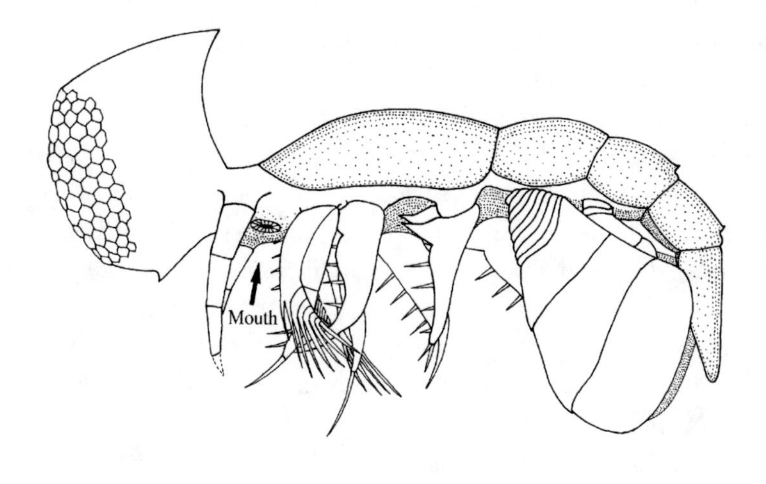

Figure 7.9 The tiny Cambrian arthropod *Cambropachycope*, with a single compound eye.

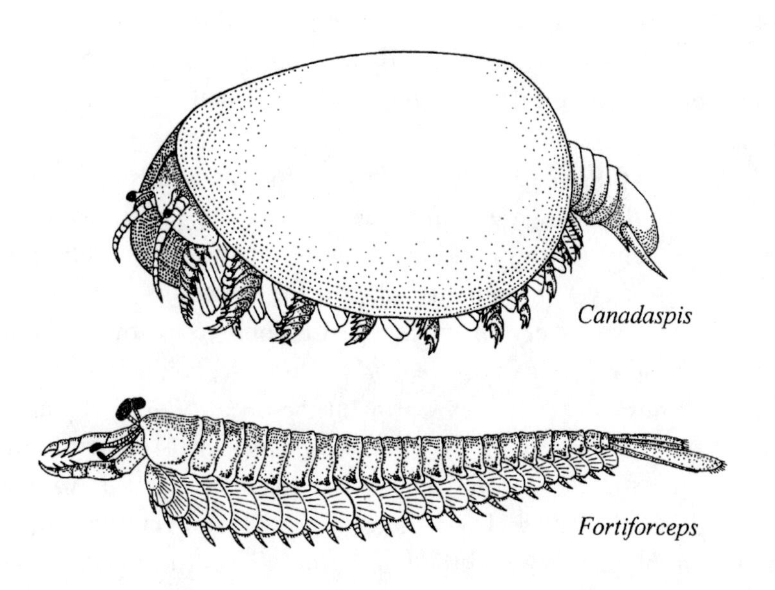

Figure 7.10 The Cambrian arthropods *Canadaspis laevigata* and *Fortiforceps foliosa* from Chengjiang, China.

exceptionally well-preserved suite of Cambrian fossils has been recovered. And these 'Chengjiang' fossils are ten million years older than those of the Burgess Shale.

The Chengjiang fossils also include many species with eyes. Here there are both stalked eyes that are moveable, and sessile eyes that are fused with the body in any of several possible positions. They can arise from the underside of the animal, extending forwards under the front margin of the head shield, such as in *Fuxianhuia*, *Leanchoilia* and *Isoxys*. In *Retifacies*, however, the eyes also sprout from the underside of the body but do not protrude forward. Then again the eyes of Chengjiang animals can also be positioned on top of the body, such as in *Xandarella*.

Like the Burgess fossils, most if not all eyes in the Chengjiang community belong to the arthropods. And these two fossil assemblages have been used to trace the position of the eye on the body through time. It is believed that the compound eyes of Cambrian arthropods shifted position from the under side to the top side of the body, and became successively incorporated into the shield or shell that covers the head. I am not sure how much we can read into this, but the same event may have taken place independently in another group of arthropods to those considered up till now – the trilobites. I will return to trilobites shortly.

It is interesting that nearly all of the Cambrian animals I have mentioned up to this point are arthropods – they belong to the phylum with hard, external skeletons that include crabs and insects. But during my description of eyes in animals today in the first half of this chapter, other phyla were very much involved. There were the swimming alciopid bristle worms in the *annelid* phylum (1), box jellyfishes in the *cnidarian* phylum (2), velvet worms in the *onycophoran* phylum (3), cuttlefishes and snails in the *molluscan* phylum (4), and of course ourselves in the *chordate* phylum (5). These animals all have image-forming 'simple' eyes. Then there were the ark clams within the *molluscan* phylum (4), and the fan worms within the *annelid* phylum (1) that, along with the arthropods (6), possess image-forming 'compound' eyes. But do any of these non-arthropod animals have Cambrian ancestors with eyes?

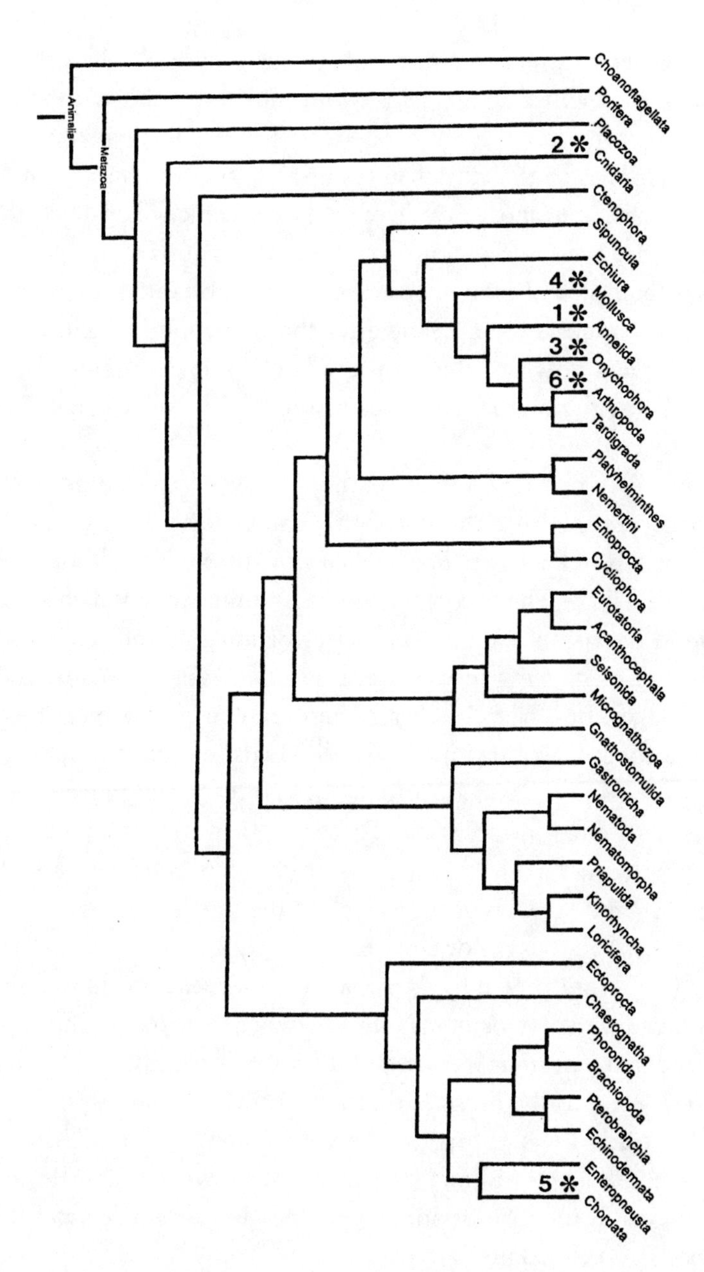

Figure 7.11 The evolutionary tree of animals at the level of phyla (all those with representatives alive today are included; note that Choanoflagellata is not a true multicelled group). Asterisks mark the phyla with eyes (which are also numbered 1 to 6 as they appear in the text). Modified from a paper by Rouse and Fauchald.

The answer to this question can obviously be 'no' when the eyed group did not evolve within its phylum until after the Cambrian, as determined from computer-generated predictions of the evolutionary tree. This applies to the cuttlefish group, a branch within the mollusc phylum. In fact the most primitive molluscs, which date back to the Cambrian, are eyeless. For similar reasons, the groups of bristle worms with eyes today can also be ruled out of the Cambrian eye club. So who is left in this ancient visual circle after the first round of elimination? The contenders now are the arthropods (1) and chordates (2), who together boast the majority of eyes today, and velvet worms (3) and box jellyfish (4).

The box jellyfish and velvet worms can be discarded as hosts of Cambrian eyes, because they probably can't see as such today. Both groups probably cannot see images that flow through the brain like the frames of a movie. The box jellyfish has no brain with which to interpret the information coding for a series of images, and its single eye remains very much a mystery. The eyes of velvet worms do not produce proper images, but are probably rather adapted to movement – they notice the approach of fast-moving individuals, but cannot make identifications. These organs may virtually bypass the brain. In true eyes, an image is assembled in the brain. The brain then makes a decision on how to react, and has the whole body at its disposal. In the case of the box jellyfish and velvet worms, as well as the bristle worms with compound eyes, their 'visual organs' may simply be binary detectors. A visual signal is interpreted by the organ as either react or do nothing. A camouflaged velvet worm may freeze when a fast-moving animal approaches. The brain is not needed during this process – the detector is wired directly to the muscles that perform the single response. This form of detection has nothing to do with vision. And to substantiate this further, the fossils of box jellyfish and velvet worms provide no evidence of eyes in the Cambrian. So now our list is reduced to just the arthropods and chordates.

Sometimes relatives of today's eyed species did exist in the Cambrian. To decide whether these ancestors, or indeed any extinct group, possessed eyes in the Cambrian, we must turn to Cambrian fossils and the law of minimum eye size.

There are few chordates known from the Cambrian. The best known are *Pikaia* from the Burgess Shale, and the earliest of all *Haikouella* from Chengjiang. Fossils of *Pikaia* reveal a clear body outline along with fine details of internal parts, including muscles and a notochord, a kind of backbone. But features of the front end of the animal are too small to be seen without a microscope. They are, consequently, too small to be eyes. The same can be said of all Cambrian chordates – they could not see.

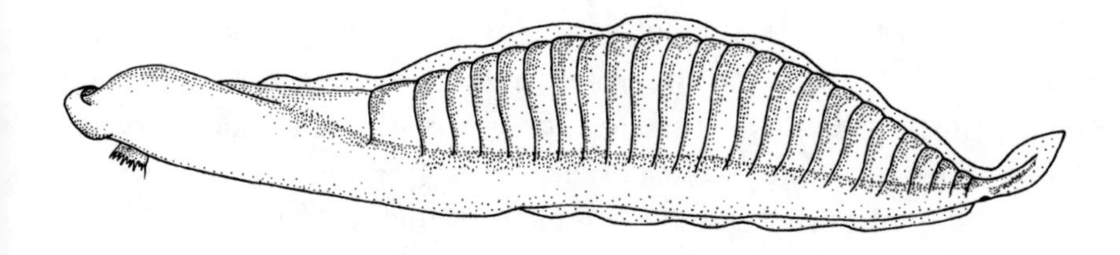

Figure 7.12 *Haikouella lanceolata* from Chengjiang – the earliest known chordate.

Today most blind chordates live in environments with extremely little or no light. Think of the mole. And then there is the Mexican cave fish that has eyes where light is present, and no eyes where light is absent. But there are at least two species of chordate known from the Cambrian, and they lived in sunlit environments. Indeed, many of their neighbours had eyes. So where most of this group have eyes today, why did they not in the Cambrian? This is not what we would expect. The idea that life happened in the Cambrian as it continues today fits for arthropods – they can see now and they could see then. And most modern chordates can see.

So far I have considered that the eye evolved at only one point in time, and that all eyes in existence today stem from that ancestral organ. This implies that the eye must have evolved before the divergence on the evolutionary tree of all animals with eyes. The animal with the ancestral eye must have been the ancestor to the arthropods, chordates, bristle worms and molluscs – animals with eyes today. In which case the eye must have evolved hundreds of millions of years before the Cambrian explosion, when these phyla diverged from each other (albeit remaining within similar, soft bodies). Things, however, did not happen this way.

There *are* chordates living in sunlit environments today that have no eyes. They are the most primitive forms of chordate – the type that existed in the Cambrian. I refer to the hagfish, and animals even more primitive. If the most 'primitive' chordates did not possess eyes but the more derived chordates did, this means that the first chordate eye evolved at some point *within* the chordate branch of the evolutionary tree. And now we can justify the lack of chordate eyes in the Cambrian – the eye has a multiple origin. It evolved on more than one occasion – the arthropod eye evolved and the chordate eye evolved, but independently and, it seems, at different points in evolutionary history. When the chordates first branched out from the evolutionary tree they did not have eyes. And the same goes for all other phyla. Now it seems *more* than possible that an eye appeared on Earth in one phylum before any others – it seems veritable. And that phylum with the first eye was the Arthropoda.

There is one group of arthropods I have yet to examine, a group well represented in the Burgess Shale. These are the trilobites.

Earlier in this chapter I casually added trilobites to the list of arthropods with compound eyes. But I did not suggest the exceptional nature of the trilobite. Compound eyes were common in trilobites, which in turn were common in the Cambrian, so it is appropriate to devote part of this chapter to trilobite eyes.

We know that trilobites reigned in abundance throughout the seas. This reign ended 280 million years ago, but began 543 million years ago, at the beginning of the Cambrian explosion. Four thousand species of trilobites have been identified, and they were particularly successful during the first term of their dominion, when they flourished.

We need not rely on the Burgess Shale and Chengjiang fossils for information about trilobites in the Cambrian. Trilobites are found all over the world and from all periods within the Cambrian – their preservation was not dependent on particularly favourable conditions. And the diversity of Cambrian trilobites suggests they were by far the most important and ubiquitous arthropods around in the Cambrian. In fact trilobites are believed to be the stem group of all arthropods – they probably wore the prototype shells, or 'exoskeletons'. From some groups of trilobites the crustaceans, and later the insects, evolved. From another group the sea spiders, and later the spiders, evolved.

The exceptional preservation of trilobites can be attributed to the constituents of their shells – fossilisation-friendly chemicals. And the conservation of their optics allows us a glimpse into their vision – most trilobites had compound eyes.

The compound eyes of trilobites are different from the true compound eyes of today in that their lenses were made of the mineral calcite. Calcite is widespread on Earth – chalk is calcite, but granular, so that it appears white via scattering. Scattering causes structural colour – a white or blue appearance depending on the size of the scattering elements. The elements, or granules, are relatively large in chalk, which causes all wavelengths in white light to be reflected equally and in all directions. And, as Newton demonstrated, when all wavelengths combine the light appears white. But if the calcite is formed slowly, a perfect crystal results, completely free of granules. This type of calcite is crystal clear, and was the ingredient of trilobite lenses. Today, calcite lenses are found only in bristlestars, relatives of starfish. And these lenses are not part of an eye as such, rather a component in a composite light perceiver comparable to that of some bristle worms. Although all relied on calcite lenses, there were two distinct types of compound eyes in trilobites – holochroal and schizochroal.

Schizochroal eyes were big, but not because of the number of facets they contained, which were surprisingly sparse. Instead they owed their size to the vastness of each facet – up to a whole millimetre in diameter, a dimension not even approached in today's compound eyes. A boundary region separated each facet from its neighbours, and the lenses were either elongated prisms or came in two parts that locked together, one above the other.

The two-part lens is interesting. In addition to flying a kite into a thunderstorm in an attempt to understand electricity, the American diplomat and scientist Benjamin Franklin was also famous for inventing bifocal glasses in the eighteenth century. These offered the wearer

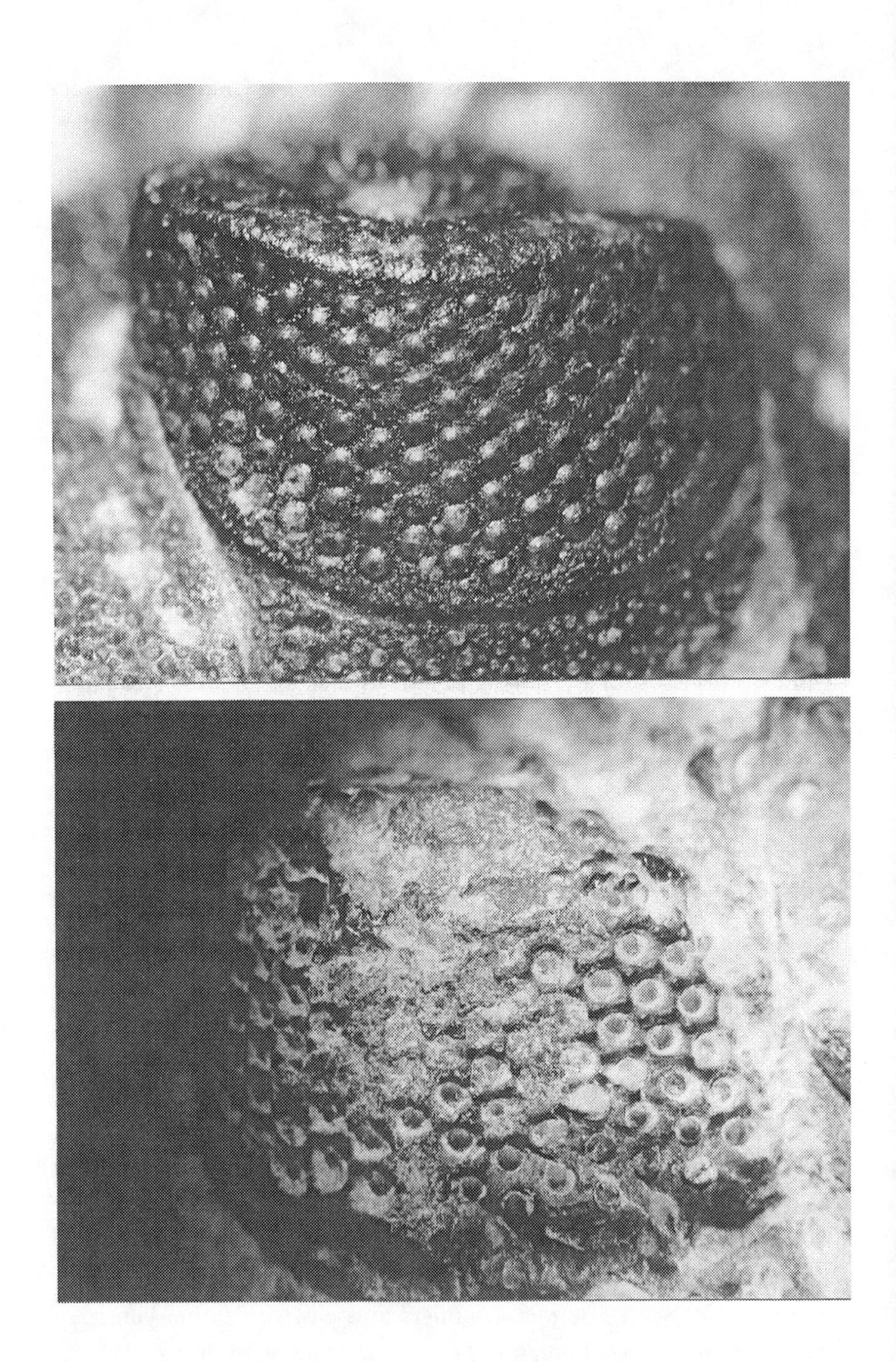

Figure 7.13 Photographs of holochroal (above) and schizochroal (below) trilobite eyes.

the choice of seeing objects either near or far with accuracy. The trilobite schizochroal eyes with the two-part lenses did the same thing, so that they could see both tiny prey within grasping range and enemies approaching from a safe distance. The graded material lenses first identified by Maxwell in his breakfast herring are also to be found in some schizochroal eyes. Making a direct comparison, these trilobite eyes may have worked like those superposition eyes with graded lenses, such as are found in a group of crustaceans called mysids. On top of this, a new type of eye was found recently in a living insect called a strepsipteran, and this may unlock the secrets of the enormousness of the schizochroal facets.

Strepsipterans are tiny insects that parasitise wasps. The eye of the male strepsipteran has only fifty lenses, as opposed to seven hundred in the similar sized *Drosophila* fruit flies, but they are each relatively huge. Each lens is linked to its own retina, and the nerves serving the individual retinas cross over, so that a complete picture can be assembled in the brain with everything in the right place. This form of imaging, which was probably echoed in the schizochroal trilobite eye, falls somewhere between that of compound and simple eyes.

Efficient as it may seem, and strepsipterans aside, the schizochroal eye was confined to just one group of trilobites known as the Phacopina. Phacopina lived up until 370 million years ago, but they did not evolve until the very end of the Cambrian period, around 510 million years ago. The schizochroal eye evolved *from* the trilobite eye of most interest in this chapter – the holochroal type. This has a significantly earlier origin.

Holochroal eyes generally contained more facets than schizochroal eyes, where each facet was relatively small. The lenses were simple – thin and biconvex ('oval'), like the lenses of magnifying glasses. They were packed together in a square or hexagonal formation, where neighbouring lenses were touching. But exactly how the holochroal eye functioned is something of a mystery. The real problem is that part of the eye may or may not be missing in the fossils. We do not know. The calcite lenses have preserved well thanks to their chemistry. But were there further focusing elements lying behind these that just haven't preserved because of an unfavourable chemistry for fossilisation? In one way, taking a strict view of their position in the eye, the calcite lenses of trilobites could be more comparable to the thick corneas of modern compound eyes. In which case we *would* expect a further focusing element, or lens, to be lying just beneath. But then again, maybe the calcite lenses were the *only* focusing elements of holochroal eyes in trilobites, and maybe these were quite adequate.

So the internal architecture of the trilobite eye, as it is known, cannot provide the information needed to make comparisons with modern compound eyes. But there are some clues to be found on the outside. Those holochroal eyes with square-shaped facets may be comparable to the reflecting type of superposition compound eye, where the facets are square for a reason. These are the facets lined with mirrors, and the mirrors perform the focusing. And then those holochroal eyes with hexagonal-shaped facets may have worked like today's apposition compound eyes, with stark similarities on the outside. If these inferences are correct, we could predict the environments or lifestyles of the trilobites.

One shrimp today changes its eye throughout its development from juvenile to adult. As a juvenile it has an eye with hexagonal facets – an adaptation to its bright, shallow-water environment. This apposition eye is good at producing sharp images, but not so good at collecting all the light available. Fortunately there is a plentiful supply of light for this juvenile. But as it grows, it migrates to deeper waters, where light becomes more limited. So the apposition eye is shed during the moult to adulthood, and is replaced by a superposition eye with square, mirror-box lenses. This new eye has quite the opposite properties; although not effective at forming sharp images, it can make the most of the light available. All of this evidence suggests that trilobites with hexagonal-shaped facets lived in shallow waters, and those with square-shaped facets lived deeper, or were active at night.

On the other hand, square and hexagonal shaped corneas may be consequences of lens-packing geometry alone – the way they are squashed together. Because an alternative exists, unless completely preserved holochroal-type eyes are found in the future we will never know with certainty exactly how this organ worked and, consequently, how its hosts viewed the world. For the purposes of this book, it is enough to say with confidence that trilobite eyes produced visual images – trilobites with these eyes could see. Now we should move on to more important matters.

Although the origin of the holochroal eye *now* appears to pose a big question, it is a question that has never been properly addressed. Without the line of enquiry maintained in this chapter, there is little justification for pursuing such an otherwise unimportant goal within the gravity of science. But in this chapter we have been deliberately building up to the very first eye in existence on Earth. Through a process of elimination we have arrived at the holochroal compound eye of trilobites. And from here we are in the hands of palaeontologists. We must rely on the fossils to help provide us with a date for that *very first* holochroal eye – the *very first eye*. The fossils do not let us down.

The oldest trilobites known are from the Lower Cambrian – the earliest part of the Cambrian. So far, so good. But we can be even more precise than that. The very first trilobites evolved at the *very beginning* of the Cambrian, around 543 million years ago – and they were equipped with holochroal compound eyes. Before this date there were neither trilobites nor eyes on Earth. So it is worth a look at those first trilobites and their eyes.

The oldest known, *well-preserved* trilobite eyes were described by Euan Clarkson, an expert on trilobite eyes at the University of Edinburgh, and his colleague Zhang Xi-guang from the Chengdu Institute of Geology in China. Working on material from south central China, they found particularly interesting compound eyes in two species of trilobite – *Neocobboldia chinlinica* and *Shizhudiscus longquanensis*.

Xi-guang and Clarkson used acid to dissolve the limestone slabs excavated from their Lower Cambrian site. The trilobites were freed from their matrix and were ready to be studied in electron microscopes. They had been particularly well preserved, thanks to the protection afforded by a phosphate coat, and so fine details of their optics could be uncovered.

Neocobboldia bore a thick lens in each facet of its eye, and the lens was free from spherical aberration, the problem created when rays entering different parts of the lens are focused on to different planes. But it showed no signs of a graded material lens like that of the herring or some compound eyes today. How was spherical aberration avoided? The answer lay in a sophisticated design, involving a precisely curved divide within the lens. This 'intralensar bowl' design was not new to science – Huygens and Descartes had invented something similar in the seventeenth century – but here trilobites were proving that it really worked.

The lenses are less well preserved in the eyes of *Shizhudiscus*, although they are obviously simpler in design – they are biconvex ('oval'). All things considered, these eyes conform to the rules of the holochroal type and so they could produce visual images in the very

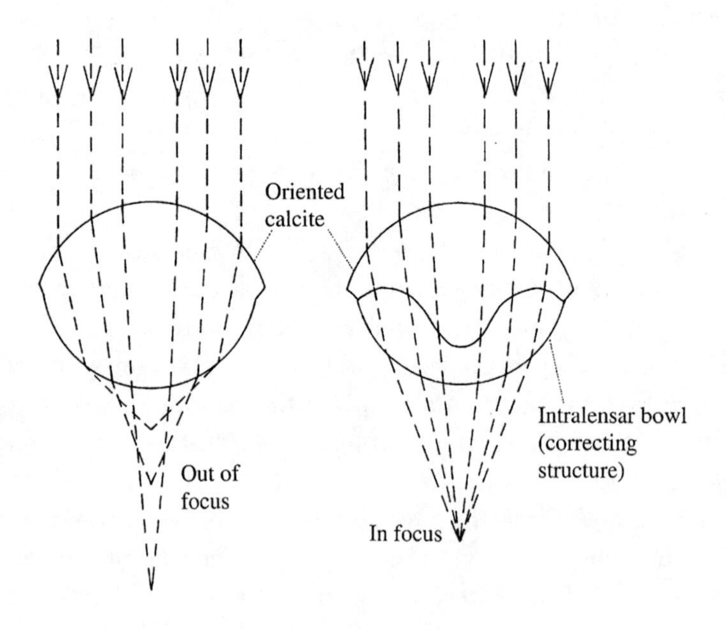

Figure 7.14 The intralensar bowl design in the lenses of some trilobites; light rays striking all parts of the lens are focused in the same plane. An identically shaped lens without the intralensar bowl is shown for comparison.

Early Cambrian. And nice panoramic pictures, too – images of anything positioned on the trilobites' horizon.

In fact there are a number of trilobite species known from the very

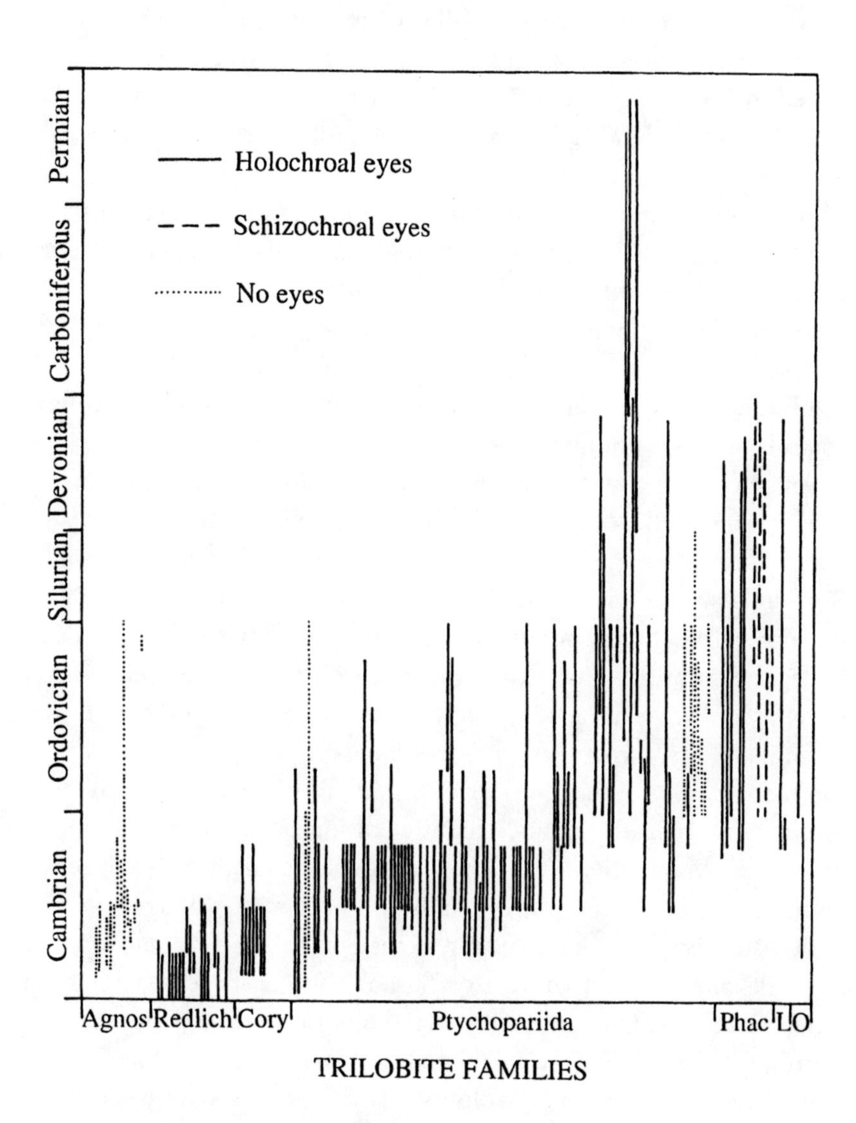

Figure 7.15 Time ranges of genera within the seven families of trilobite, showing the occurrence of different kinds of eye (after Euan Clarkson, 1973). Note that the very first trilobites, living at the base of the Cambrian, bore (holochroal) eyes.

beginning of the Cambrian, and all have eyes, although they are not so well preserved. Even Charles Doolittle Walcott suspected this trend in 1910. And in 1957 the trilobite specialist Frank Raw, of Birmingham University in England, declared, 'How ancient already in the Lower Cambrian must the compound eye have been.' This statement was seconded by Euan Clarkson in 1973. Another case of trilobite eyes is that of *Fallotaspis* from Morocco, around 540 million years old. *Fallotaspis* had large eyes. The list continues . . .

To reiterate, there is one curious but common fact emerging from the trilobite eye data. Many species of trilobites with eyes came into existence around 543 million years ago . . . but not a single species before that time. The trilobites *without* eyes entered history a little later, in geological time. So 543 million years ago the Earth witnessed the first trilobite . . . and the first eye. Five hundred and forty-three million could be a magic number.

An important question in this study is, 'How quickly can an eye evolve from its forebears?' Fossil evidence implies that eyes existed 543 million years ago but not before. Not, say, 544 million years ago. But surely an eye cannot simply evolve overnight? Surely it has to pass through a sequence of intermediate stages, probably within intermediate species in the evolutionary tree? These intermediate species must have fallen between the completely eyeless ancestor and the first eyed ascendants. If some of these intermediate species could see, albeit in a rudimentary way, and if they existed millions of years before the first ideal eye, then maybe animals acquired vision earlier than 543 million years ago. Maybe the introduction of eyes on Earth was staggered; perhaps sharp images were seen millions of years after blurred images, with visual precision increasing systematically. Dan-Eric Nilsson and his colleague at Lund University, Susanne Pelger, have deciphered a series of intermediate stages for the evolution of a camera-type eye. More than that, they have calculated the time needed to complete the sequence via the process of evolution. This is just the data we need.

At the beginning of this chapter we examined light perceivers that could detect light levels but not form visual images. They were not eyes. But some light perceivers were more efficient than others, and it is likely that a more efficient type originated from a less efficient type during its evolution. To make their predictions, Nilsson and Pelger applied this logic.

A patch of light-sensitive skin was used as a starting point. This dents inwards, and becomes increasingly infolded to form detectors escalating in their sensitivity to the direction of light. This assumption is quite acceptable since all the intermediate stages can be found functioning in animals today. It is important that each link in the chain *can* exist in its own right. The opposite of this was once used to criticise evolution, and even clouded the thoughts of Darwin himself, as suggested

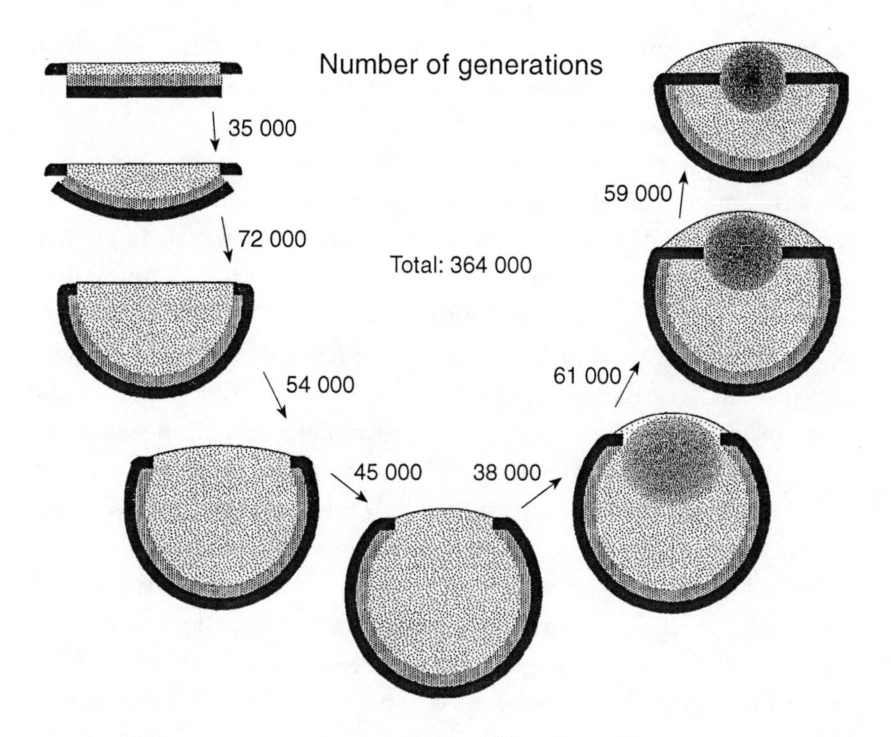

Figure 7.16 Nilsson and Pelger's predicted evolution of a camera-type eye, like that of a fish. The sequence begins with a flat patch of light-sensitive cells sandwiched between a transparent protective layer and a layer of dark pigment. A graded-index lens appears at stage 6. Reproduced from a 1994 paper by Nilsson and Pelger with permission from the authors.

in the epigraph at the beginning of this chapter. To justify this further, we can explain why all animals don't possess the theoretical ultimate eye. The intermediate stages, or conceptually substandard visual organs, *do* exist today because their host animals cannot handle the information loads supplied by the next conceivable stage on the road to a fully formed camera-type eye – Darwin had no reason to be concerned. Back on the evolutionary road, we have reached a 'cup eye' that cannot form proper images. We have also reached a junction. Close the entrance to the cup even more and we have the pinhole eye of nautilus. Then again, begin to grow a lens and another path has been taken – the path to the camera-type eye typical of vertebrates.

Nilsson and Pelger were more than realistic in assuming that a light receptor will change by just 1 per cent of its length, width or protein density during each evolutionary step in the eye direction. But even with such a pessimistic approach, the whole sequence from light-sensitive patch to the eye of a fish would require only two thousand of these tiny modifications in sequence. That may not seem enough, but as Michael Land and Dan-Eric Nilsson point out, if two thousand sequential modifications of 1 per cent are applied to the length of a finger, then it becomes long enough to bridge the Atlantic Ocean.

We know that proteins need not evolve from their chemical beginnings. A study of flatworms revealed that similar proteins exist in the eyespots (not true eyes) *and* touch/chemical detectors. In the eyespots, these are the proteins that react to light, comparable to those in the retina of an eye. So a head start may be gained towards eye evolution by borrowing the proteins of other detectors.

Now for the calculation of time needed for these modifications to take place, which is really what we are interested in. Again, caution was the name of the game when Nilsson and Pelger made their assumption about the slowest rate of evolution – a 0.005 per cent modification from one generation to the next. In reality, the rate would probably be faster. For instance, the light receptor pigments of modern crustaceans show an evolution that is considerably more *rapid* than expected. And verily the word 'pessimistic' entered the title of Nilsson and Pelger's original paper, which made their result seem even more remarkable. They found that the eye of a fish could evolve from its rudimentary beginnings in less than 400,000 generations. Assuming each generation is completed within a year, this result suggests that an efficient, image-forming eye can evolve in less than half a million years. Now that really is a blink of an eye on the geological timescale.

This is a camera-type eye and we have established that the first eye was compound. But in their definitive book on the optics of animal eyes, called *Animal Eyes*, Michael Land and Dan-Eric Nilsson were beginning to picture the evolutionary sequence of the compound eye. They claimed that arthropods 'probably originated from a worm-like ancestor that already possessed a rudimentary compound eye – possibly a loose collection of eyespots'. Independently, the Australian biologist Richard Smith mapped the changes needed to form the compound 'eye' of a bristle worm. A loose collection of eyespots also appeared in Smith's sequence. And the number of links expected in the chain leading to a fully functioning eye was on a par with those in Nilsson and Pelger's predictions for the camera-type eye.

Like the proteins of the retina, other parts of the body involved in the process of light perception seem quite accommodating to these calculations on the eyes themselves. Nilsson and Pelger's time prediction would be meaningless if development of the visual processing centre in the brain was lagging behind that of the eye. In 1959 the biologist von Bekesy demonstrated that the effects caused by sound can be mimicked by vibrating the skin. This demonstrated that the ear and skin shared certain common features, namely nerves, in the processing of sensory information. But what does this mean for the evolution of the eye? Well, it is conceivable that nerves used by one sense can be 'upgraded' for use by two senses. And if the senses of hearing and touch can share features, then so might vision and touch. In this way, the nerves needed to service an eye would not have to evolve from a vestigial beginning – they would have a head start. Then there is a possible helping hand in the brain department. Parts of the brain, it appears, may be capable of converting from touch to vision. Dan-Eric Nilsson suggested that the compound 'eye' of ark clams and bristle worms evolved from chemical detectors that were inhibited by light. So the evolution of the *eye itself* appears to be the limiting factor, or at the back of the pack, on the evolutionary road to vision – the remainder of the system can simply be adopted. Indeed, there were other sense organs surrounding the eyes of trilobites, and the original light perceivers may have borrowed nerves from these.

Now we can calm our own nerves that may have been jangling while we gave the compound eye just one million years to evolve – at least if it was to fit with our fossil evidence. It seems that our demand has been met – one million years is plenty of time for an eye to evolve. Now we can paint a picture of 544 million years ago, where light sensitive patches were evident in the ancestors of the Cambrian trilobites. Then we can paint another picture of 543 million years ago, just the other side of the Cambrian border, where a trilobite proudly flaunts its eyes. Between the two pictures the light-sensitive patch had evolved into an eye.

Between 544 and 543 million years ago a revolution took place. During this one million year period, vision was born.

We are now in a position to interpret the statement 'How ancient already in the Lower Cambrian must the compound eye have been' made by Frank Raw. Yes, the compound eye and vision were well developed in the Lower Cambrian. But no, it was not ancient – it was contemporary. And it became the new fashion.

There was always going to be one moment in history when the eye suddenly appeared on Earth, as if out of nowhere. Now we can identify that moment. And a really important point to bear in mind at all times is that light-sensitive patches and other stages of rudimentary light receptors are not eyes. While only these patches existed, when eyes were awaiting their introduction to Earth, there was no such thing as vision.

We now know that eyes existed at the very beginning of the Cambrian . . . but not before. These two facts could be as important as each other. Considered together, they describe the introduction of a sense. Not just any old sense, but the most powerful sense or stimulus to animal behaviour and evolution in sunlit environments. And a sunlit environment is where the Burgess Shale and other well-known Cambrian animals lived. It also played host to the Cambrian explosion.

Extrapolating further, there are lifestyles that can be reconstructed based on the optics of eyes. The architecture of eyes alone can provide information on how animals lived. For instance, the position of the eyes in the head can reveal the position of the animal in the food chain. Eyes positioned at the sides of the head, facing sideways like those of a rabbit, can scan a wide angle and spot movement from nearly all directions. The movement pursued in this case is that of predators – eyes of this type belong mainly to plant eaters. In contrast, eyes positioned together at the front of the head, facing forward like those of an owl, see less of the environment but are better for pinpointing targets and judging the distance between them. These eyes generally belong to meat eaters. But this is a theme for another chapter.

8

The Killer Instinct

A little alarm now and then keeps life from stagnation

F. BURNEY (Mme d'Arbley), *Camilla* (1796)

THE LAWS OF LIFE

For the survival of animals everywhere

CONTENTS

Basic Rules

1. *Every man for himself: stay alive!*
1a. Avoid being eaten
1b. 'Eat'

2. *For the good of one's kind.*
2a. Breed
2b. Find a niche and protect it
2c. Adapt to changes in the environment

Lifestyle

1. Predator
2. Prey

Tactics

1. Conspicuousness
2. Crypsis/illusiveness
3. Genuine strength/ability

The previous chapter could be viewed as 'end of story'. Certainly, there is considerable evidence within that chapter suitable for the Cambrian files. But it is too early to jump to conclusions just yet, for there is something else to consider, a subject that has raised its head, either plainly or rather more cryptically, in every chapter so far. In each case it merged into the background as quickly as it appeared. Before ending our Cambrian investigation, we should introduce *predators* into the evidence.

The first rule of animal survival is to stay alive. The other rules, such as feeding and breeding, are academic if this first rule is not followed. But from the beginning we must distinguish between an individual and a species. A species is a collection of like individuals, which interbreed in their natural environment. Staying alive and feeding are factors that directly affect individuals, then indirectly species. Breeding and niche occupations are concerns for the long-term survival of the species. Of course animals don't really receive rules – in reality the rules for their survival are the selective pressures for evolution, invisible forces acting on the genes, carrying messages for enhanced survival. And selective pressures act directly upon individuals, not species, so even the species-level survival factors are relayed through individuals.

The first basic rule of species survival – for individuals to stay alive – will form the subject of this chapter. And more specifically, I will centre on the most important aspect of that rule, to avoid being eaten. This chapter is a stage for the predators. And, in keeping with the previous chapters, the stage will have a space *and* a time dimension.

Before launching into the world of *T. rex* and the like, I will make a brief disclaimer relating to The Laws of Life outlined on the previous page. These are the general rules but do not cover all possibilities, particularly those less common natural catastrophes. Some things are beyond evolution, such as meteor impacts, sudden ice ages, and disease. Disease is density dependent, and so it is a factor operating at the species level. On the one hand, species can become too successful for their own good. From another viewpoint, this is just evolution maintaining biodiversity, preventing one species from taking over the world. But in general, biodiversity is maintained by *all* branches of the evolutionary tree adhering to The Laws of Life. A predator does not become an overnight success by growing bigger teeth. The other side of the coin is the 'Avoid being eaten' alarm for its prey species, which favours genetic mutations for stronger armour. Cichlid fishes feed on snails, and where the fishes evolve stronger teeth, the snails simply evolve harder shells. Evolution can take animals down different roads. There are roads to predation and there are roads to prey, with the predator and prey roads running between. But all roads are endless, and animals are continuously moving along *all* of them. However, all animals today are travelling along an established evolutionary road – snails already possess armour that may yet become reinforced.

Central to this book so far have been the subjects of light and vision. When superimposed on to The Laws of Life, their capacity will become evident. Specifically, they fall into the 'Tactics' section. Consider the Hawaiian unicorn fish with its conspicuous yellow spine near its tail. The spine serves to protect the fish from predators and competitors, and consequently the unicorn fish avoids being eaten and protects its niche. But the unicorn fish rarely calls upon its spine because in reality this armament is only an ornament. Here the messenger is light. Potential predators and competitors see the armoury and have second thoughts.

When adaptations to vision include shape and behaviour, in addition to colour, it is clear that vision is a major tactic used in the struggle for both conspicuousness and illusiveness. Genuine strength or ability is actually a rare attribute in animals; rarely does an animal dominate an ecosystem without considerable employment of warnings or illusions. The lioness is the main predator in the Serengeti, but she cannot outrun her prey over short *and* long distances, so she must rely on camouflage colours and stalking behaviour to take up a competitive position in her race for food. An exception here can be found in many birds, and the reason for this exception will offer another clue towards solving the Cambrian enigma. Birds will be considered in the following chapter.

There are tactics animals can use other than vision to achieve either conspicuousness or illusiveness; other senses do exist, as also described earlier. Once again, adaptation to light is generally the main tactic to employ within The Laws of Life because of the factor separating light from all other stimuli – occurrence. Light exists, like it or not. Add Chapter 7 to the mixture and we have 'Vision exists, like it or not'. Over 95 per cent of all multicelled animals today have eyes, so if one of them is to avoid being eaten, it must be adapted to the light in its environment. We are beginning to take our knowledge of light and vision into the subject of predation.

Another thing about eyes

Chapter 7 centred on the optics of eyes, the equipment that forms an image on a retina. The reason for this was the link between the living and extinct – the optical origins of today's eyes can be traced in the fossil record, right back to the very first eyes in the Cambrian. But there is something else we can learn from the type of image formed in the past, or the view of the world through fossil eyes, that is relevant to this chapter. Just as we did in Chapter 7, we must first look for evidence in the present day.

We have learnt that there are alternative ways of producing an image today – different types of eyes do exist. But that is not the end of the variation. There are also different ways in which eyes can be arranged in a head, and these provide different views of the world.

Among the vertebrates within the chordate phylum only camera-type eyes exist. In humans they lie next to each other in the front plane of the head – they face forward. But more than that, they always focus on the same object. So why bother with two eyes, when one would appear to do the job of seeing on its own? Has evolution been excessive in our case?

When eyes are positioned on the sides of the head, like those of rabbits, the wide field of view encapsulates almost the entire horizon. At first this would seem like the ideal form of vision, but to gain such a panoramic outlook, each eye sees a different picture – each approaching 180° of the horizon – and never the same object. With one eye, however, the view will be two-dimensional, and so distances are difficult to estimate.

When two eyes are positioned on the front of the head, distances and the direction in which one is travelling *can* be estimated. So it follows that eyes in this arrangement can perceive the three-dimensionality of an object. Differences in the positions of images create impressions of depth, as can be demonstrated using stereograms. Each eye sees the same object but from a different angle. Stereograms probably work because the optic nerves serving slightly different regions of the two retinas converge on the same 'binocular' cell in the brain. The view of an object from two different angles is superimposed and averaged – and its depth is perceived. So animals with two eyes facing forward are said to have stereoscopic vision – they can perceive images in 3D.

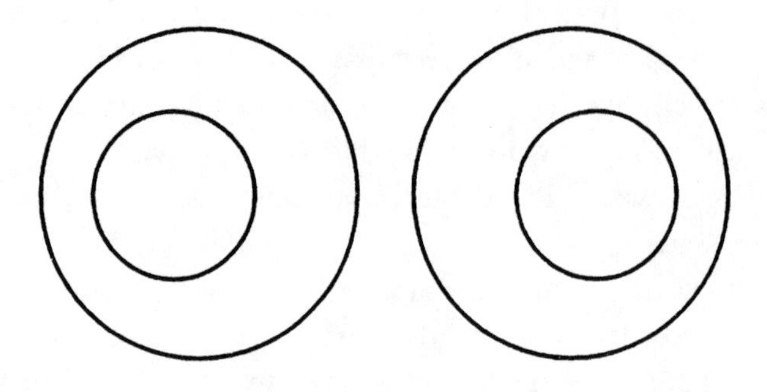

Figure 8.1 One of the original stereograms of 1838. Blur the picture to produce a fused image in the centre. The inner ring will appear nearer than the outer ring.

The stereogram, albeit merely a demonstrative game, would not work for a rabbit, or for ourselves if we closed one eye. So we should again consider whether it is better to have two eyes at the front of our head, facing forwards, or on the sides of the head, giving a panoramic view. The answer would appear to depend on the purpose of vision. Would you like to observe events happening all around you in two dimensions? Or would you prefer to view objects in front of you in three dimensions and with information on distances? Now we can return to The Laws of Life, and consider whether you are a predator or prey.

For a prey species, staying alive first means keeping off the dinner plate and *then* eating becomes important. So it is ideal for the prey species to be surrounded by open space, where the possibility of a sudden ambush is minimised. Minimised, that is, if a 360° view of the terrain is possible – blind spots on the horizon are dangerous. We often find rabbits grazing in the middle of open fields rather than at its edges near hedgerows. And we always find them with their eyes positioned for a panoramic view: eyes positioned at the sides of the head are good for spotting predators.

For a predator, in contrast, staying alive usually means eating first and worrying about *their* predators and competitors after that. Eating lively animals involves hunting. Estimating distances is a critical part of hunting – the lioness cannot begin her charge when the prey is within its safety zone, where its head start is insurmountable for the lioness. Equally, a fox cannot catch a rabbit if the rabbit is given the distance in which to reach full speed. So where vision is the major sense employed by predators, two eyes at the front of the head are needed – an accurate assessment of distances is the difference between a meal and hunger. And that is just what is found in the lioness and the fox.

This trend can often be found within other animal phyla with eyes. But in mid-water, things become more complicated. There is not only the horizon to worry about, there is also above and below. In mid-water, danger can approach from *all* directions. The great bearers of marine compound eyes, the crustaceans, have evolved a solution to this problem – many crustaceans have eyes positioned at the ends of moveable stalks. They can move their precision eyes to cover a wide area of their surroundings. Because of this, stalked eyes generally do not provide clues as to predator or prey, although many crustaceans, like insects on land, are often both. Today they lie somewhere in the middle of the food web where avoiding predation is finely balanced with the need to eat. Other types of compound eyes, however, are more obliging to the Cambrian detective.

Later in this chapter, I will attempt to relate the feeding information provided by eyes to the inhabitants of the Cambrian. Eye stalks in this respect are like gloves to the fingerprint detective – they mask potentially useful information. But compound eyes that are fixed in position do offer some clues, and such eyes are found commonly in the fossil record.

In the air, dragonflies are expert hunters. They have three pairs of grasping limbs positioned near to their blade-like mouthparts, and large wings to provide speed and manoeuvrability. But first the helpless prey must be found, identified as prey, and then tracked. This is achieved using vision – huge eyes are fused to the head. These eyes lock the prey in their sights, their 'sights' being just parts of the eyes and not all the facets. This is food for palaeontological thought.

The compound eyes of dragonflies contain several hundred or even thousand facets, not all of which are equal. There are one or two regions of the eye that contain larger facets and these are known as the acute zones, the 'sights'. Larger facets provide higher magnification and better resolution – they see with greater sensitivity. One acute zone is positioned at the top of the eye, and this is used to scan through the air and identify prey insects against the sky. When a prey insect has been spotted, the dragonfly moves into its horizontal plane and tracks it with a forward facing acute zone – the prey is now locked into a line of fire. But the relevant point here is that the size and positions of the facets within the eye provide information on feeding – predation in this case. The eyes of prey can be quite different.

For animals that require vision only to avoid being eaten, having two eyes is just one solution. Rather than evolving a pair of good image-forming eyes capable of scanning the entire environment, they may evolve numerous, less efficient eyes distributed over a large area of the body. At the sacrifice of images, numerous eyes are ideal for detecting movement – as an object passes over them, its moving shadow is detected. When the environmental light changes, as when a fish passes through the ocean, a response is triggered. Numerous compound eyes are indeed found in nature. They occur in ark clams (molluscs) and fan worms (bristle worms) where they are employed to detect predators.

The real advantage of this multiple eye system probably lies in the word 'evolve' used at the beginning of the previous paragraph. Evolution involves changes, changes from one structure to another, for instance. Here we are back to Darwin's original doubts caused by the eye – from what could our highly complex and specialised eyes have evolved? We now know that skin and ears can share nerves, and that part of the animal brain may have converted from touch to vision at some stage. Dan-Eric Nilsson suggests that the light detector cells in the compound eyes of ark clams and fan worms evolved from chemical detector cells that were inhibited by light. Originally, these chemical detectors were distributed over a large area of the body and, consequently, so too are the eyes today. In other words, it was most accommodating in these cases to evolve eyes all over the body.

Ark clams and fan worms are preyed upon by fishes. They have soft parts used for feeding, which can be enclosed within hard parts in the form of a shell and tube respectively. So these animals would benefit from a burglar alarm, an early warning system to detect a predator's approach. And that is the function of their eyes. When the movement detected in the water equates to that of a fish, the ark clam closes its shell tight and the fan worm withdraws into its tube. The armoured doors are closed. And their many compound eyes were the cheapest evolutionary option capable of performing this function from the building materials or starting points available.

Clearly, signs are appearing that the architecture and position of eyes can reveal not only how an animal sees but also its position in the food web – whether it is a predator or prey. Chapter 7 used fossil eye architecture to trace vision in the past. Now I will re-examine the fossil evidence, where appropriate, and use it in an attempt to trace the history of predation.

The Cambrian arthropod *Cambropachycope* had a single compound eye. Other than the weird *Opabinia*, the failed five-eyed experiment, all other Cambrian eyes producing good images and with the potential for image analysis were paired. When cross-sectioned, each of *Opabinia*'s five eyes revealed the general architecture of a compound eye. But *Opabinia* had a flexible tube-like mouthpart extending from its head and terminating in a grasping jaw. The arrangement of the eyes at the front, side and top of the head are not so easy to interpret because of that mouthpart – it could extend in front of, to the sides or above the head. So which direction 'faces forwards' for *Opabinia*? Because there are several 'forward' directions for the mouth, we cannot say whether *Opabinia*'s eyes served to view the entire environment or to centre on just one direction. Before tackling the remainder of the Burgess fossils, first we must reassess *Cambropachycope*.

Cambropachycope was an ancestor of the crustaceans. Although only a few millimetres in size, it is known in great detail from a fossil site in Sweden thanks to very favourable preservation conditions. As mentioned in Chapter 7, the bulbous front end of *Cambropachycope* was an eye – a single, large compound eye. An examination of the cornea of this eye revealed that it completely covered the slightly flattened front surface of the animal. Facets were evident on the surface as it curved away towards the sides, but generally the sides were bare. Importantly, the facets on the curved edges were small compared to those of the central part of the eye. It seems that the centre of the eye saw with the greatest precision. *Cambropachycope*'s eye could scan a 120° sector of the environment – that sector in front of it. And just like in dragonflies of today, the central region of the eye could achieve finer resolution. In conclusion, this was the eye of a predator. *Cambropachycope* would have terrorised the tiny inhabitants of the Cambrian around 510 million years ago.

Unfortunately the eyes of the Burgess Shale animals do not reveal enough information on their optics to allow us to draw conclusions on feeding from just a single eye. We cannot resolve details of their individual facets. To add to this, most nontrilobite eyes in the Burgess Shale are stalked, so their manoeuvrability makes directional predictions difficult. But some are obliging to the palaeontologist.

Due to the short length of their stalks, the eyes of the Burgess arthropod *Sanctacaris* are greatly restricted in that they can only be directed forwards, suggesting a predatory lifestyle. And then another Burgess arthropod, *Yohoia*, has eyes fixed in position with bulbous, enlarged regions directed forwards, again further suggesting that predators were in existence 515 million years ago. There are other signs of predation in the Burgess Shale fossils, which will form the subject of the next part of this chapter, but first we could consider the Cambrian trilobites, which often show details of the individual facets of their compound eyes.

Most trilobite eyes, particularly the holochroal eyes that were the first to appear on Earth, have larger facets at their centres than at their edges. The eyes of early trilobites were positioned on the sides of the head, but were curved to scan the complete horizon around them. So they saw with greater precision towards their sides, at right angles to the forward direction of the trilobite. These characters appear contra-

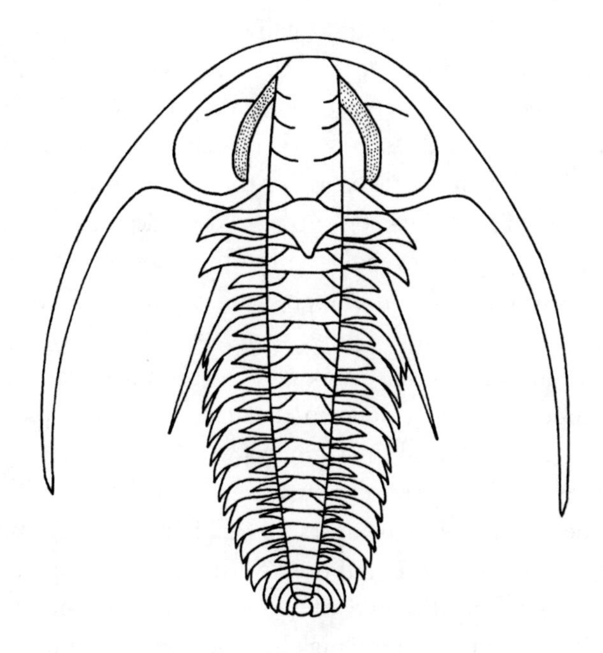

Figure 8.2 The early Cambrian trilobite *Fallotaspis typica* showing eyes (shaded) positioned at the side of the head, although its sight is directed slightly forward.

dictory when compared with most living animals – eyes on the sides of the head suggest prey, while larger facets in the centre of the eye suggest predator. But there are animals in the sea today that do have eyes with similar directional properties to the trilobites – the fishes.

Fishes have eyes positioned at the sides of their head yet do not see equally in all directions. But fishes have camera-type and not compound eyes, so how can we infer such information when there is only one lens? The answer lies with the retina, and the distribution of the light detection cells within it.

If the eyeball of a fish is cut along its 'equator', and the lower hemisphere is cut along its 'lines of longitude', it can be laid flat. A globe is viewed on the flat pages of an atlas in this manner. The lower hemisphere of the eyeball is the retina, the region of light detection cells where the image of the eye is formed. Objects to the side of the eye are imaged at the edges of the retina, while objects positioned along the central axis of the eye are imaged in the centre of the retina. The retina can be examined under microscopes to map the light detection cells. The results are always the same – the greatest concentration of light detection cells lies around the centre of the retina. Fishes see best along the central axis of their eyes, or out from the sides of the head.

Fishes can move their eyeballs within their sockets to some extent, but in comparison the region of high visual sensitivity in trilobites was larger than in fish, and so eye movements were probably not so necessary for trilobites to precisely track an object in the water. Considering that life supposedly functioned in the past as it does today, maybe trilobites were the fishes of the Cambrian.

These generalisations are obviously very broad. And fishes of open water can be predators, scavengers or herbivores, not to mention the fact that most are also potential prey. So this line of enquiry is, unfortunately, approaching the end of its usefulness, although it will be considered further in the following chapter. The ambiguity of eye position and architecture in relation to position in the food web means that we must look elsewhere for signs of predation in the Cambrian. And the most obvious place to look will turn out to be the best one – the whole fossilised bodies of the Cambrian animals themselves.

Swords, shields and scars

Post-Cambrian – the potential

So far I have been looking for secondary signs of predation. But what about primary signs – the knife-like tools and bite marks themselves? Rather than pursuing gun sights or the criminal mind, maybe we should be scouring the fossil record for the murder weapons and victims? And what of the shields used to protect from those weapons? It is worth searching for this evidence.

One of the most interesting museum specimens must be the 'Death of a goanna', the centrepiece of the Queensland Museum in Australia. This metre-long goanna, or monitor lizard, is preserved in its death pose – mouth wide open and stuffed with an echidna. The goanna (half-wittedly) attempted to swallow the foot-long marsupial, complete with its outer coat of long spines. The spines punctured the mouth of the goanna in all directions, and the animals locked together and died in stalemate (see Plate 25).

Chicago's Field Museum of Natural History displays a comparable specimen. Here the rear half of a herring-like prey emerges from the mouth of a perch-like predator. The herring proved to be more than a mouthful as both fish died in this irreversible position. But in this case the animals inhabited a lake 50 million years ago. They are preserved as fossils in a huge slab of limestone, recovered 2,500 metres above sea level in Wyoming.

It is rare to find ancient feeding preserved in action in the fossil record, but there are primary signs of both predators and prey locked within numerous fossils. Dinosaurs provide the obvious examples. The dentition of *T. rex* can mean only one thing – it ate meat. But did it kill or poach its meat – was it a predator or scavenger? The speed at which *T. rex* ran, as calculated from footprints, suggests it was capable of catching living prey. But this question remains a little contentious.

In South Dakota, an amateur fossil hunter unearthed part of a bone from a *Hyracodon*, a thirty-million-year-old pig. Many other bones had been recovered from this extinct species, but there was something unusual about this one. Something was not quite right.

The bone was about the size of a golf ball. It was quite unexciting, except for some marks – clear, neat indentations up to a centimetre deep in places. Then the fossil hunter found a jaw from *Hoplophoneus*, a cat that lived in the same region as *Hyracodon*. The cat's dentition precisely matched the marks in the pig bone. It seemed conclusive that the cat had eaten the pig thirty million years ago. But when the ancient cat dined, was the pig already dead, or did the cat kill it? We will never know the answer. More telling, however, are the puncture marks in ammonoids.

The now extinct ammonoids, as featured in Chapters 2 and 6, lived within a hard, spiral shell, allowing their tentacles to protrude into the water. Ammonoids probably hunted in the manner of squids and cuttlefishes today. Here tentacles grip prey with their suckers, while their beak-like mouthparts and file-like inner teeth perform the cutting and grinding. But we know something else about feeding involving ammonoids – this time as prey.

During their reign, ammonoids swam successfully through ancient seas. Occasionally, however, they would be seen falling through the water, plunging to the ocean floor. These ammonoids were dead, or dying . . . but isn't this theory contradictory? When ammonoids died, the gases released from their decaying bodies inflated their shells. The buoyant shells then floated to the surface and were washed ashore, where they were laid to rest in the shallow ammonoid graveyard. Yet some sinking ammonoids were heading for a deep-water grave – but why?

Sometimes ammonoid shells *are* recovered from deep-water localities. Sometimes they *did* fall to the sea floor directly below their natural waters. The fossils in this case, however, are different from those of shallow-water graves. The shells found in the unnatural, shallow-water sites are intact. The shells recovered from their owners' original localities bear puncture marks.

The puncture marks are roughly circular and the sizes of various coins. Cracks often radiate from these marks. Some shells have puncture marks that are randomly arranged, while others have marks arranged in patterns. There are two theories for the cause of the randomly arranged marks. The first is that limpets caused them.

Limpets are snails with hat-shaped shells. They graze on rocks or other hard surfaces. After grazing they often return to the same resting place, eventually forming a shallow, round depression. One idea is that the shells of ammonoids were suitably hard surfaces which ancient limpets could make their resting places. In this case, cracks radiating from the holes would be artefacts of deep burial and, consequently, high pressure. The alternative theory, however, is more dramatic, and certainly explains the regular patterns of puncture marks formed.

Mosasaurs were large, marine reptiles that lived alongside ammonoids. Their crocodile-like dentition suggests they were predators that patrolled the open water of ancient seas. But there is something else that can be inferred from their complete dentition – that they preyed upon ammonoids.

Jaws of mosasaurs can be found which explain the patterns of marks in ammonoid shells. When the shells are placed between the jaws of a certain size, the teeth fit precisely into the marks. The size and position of the mosasaur teeth within their jaws are a perfect match with the puncture marks of ammonoids. Case closed. Now we can re-reconstruct those ancient seas with ammonoids swimming, but this time with mosasaurs snapping at them.

Whether or not the randomly arranged puncture marks were made by limpets or mosasaurs, which may have taken several bites, they do explain the deep-water burial of these ammonoids. When punctured, the ammonoid shell would begin to fill with water, although the living part of the ammonoid would remain alive. As water infiltrated the otherwise gas-filled chambers of the shell, the ammonoid would become less buoyant and start to sink. Lying helpless on the sea floor or incapable of movement on its way there, the ammonoid would be susceptible to further, fatal attacks by the mosasaur. And the shell would remain below the scene of the crime, to be buried in a deep-water locality and not with those ammonoids that died a natural death, in their shallow-water graveyards. But their problems all began when their probable camouflage cover was blown – mosasaurs were visual hunters.

The 20,380-year-old Siberian mammoth first mentioned in Chapter 2 was found alone in frozen ground and French scientists have been investigating the cause of death of this specimen in the hope of explaining the extinction of mammoths in general. But maybe a single specimen will never provide the answer to this dilemma. Bones from *many* mammoths, however, were uncovered from a site in England, and it is the number of individuals that suggests the mammoth was victim of a successful hunting strategy. That, and the marks of predation.

In a burial site for ancient Britons who lived up to 50,000 years ago (ending with, apparently, King Arthur), the bones of other animals are to be found. At Wookey Hole near Glastonbury in England there is an extensive system of caves. The entrance to these caves lies at the foot of a 50-metre, vertical cliff. At the foot of this cliff there is also a small recess, protected from the often harsh climate. Within this recess, two sets of bones have been found – those of predators and those of plant eaters, or prey. The ancient predators of Wookey Hole were hyenas, and the prey were mainly mammoths. The hyena teeth fit precisely into marks in the mammoth bones, evidence that hyenas once preyed upon mammoths. But how could hyenas kill such a huge animal as a mammoth, and how did they lure the mammoths into their den, the recess in the cliff?

Mammoth bones have also been found outside the recess, at the base of the cliff, which was a likely place of death for mammoths. But the scene of the crime was probably 50 metres above. This is not the only place on Earth in which such a scenario has been uncovered, and the pattern emerging has led to a theory of how mammoths were hunted.

It is unlikely that mammoths simply wandered too close to the edges of cliffs, rather that 50,000 years ago hyenas hunted on open plains, some of which ended abruptly at cliff edges. Pursued by hyenas towards the cliffs, it is probable that on occasion a mammoth tumbled over the edge. Mammoth bones at the foot of cliffs suggest that sometimes they fell over the edge, and the piles of bones from many individuals would seem to be more than a coincidence. So the hyenas living in the den below the cliff would be ideally placed to consume the carcasses. Theoretically, this is a good hunting strategy, and one that can be deduced from the fossil record and geological formation. But again, the real evidence that ancient hyenas ate mammoths lies in the teeth marks in the bones.

Back to the Cambrian

Discussion so far has covered events that took place long after the Cambrian period, but what about the Cambrian itself? Are the equivalent of teeth and teeth marks recorded in the Cambrian fossils? We can turn to the Burgess Shale for evidence for the last time.

In the Burgess Shale are found groups of animals that exist today as predators. The jellyfish-like comb jelly *Fasciculus* would have pulsated through the shallow Cambrian seas swallowing any suitable prey in its path. The priapulid worms *Ancalagon*, *Louisella*, *Ottoia* and *Selkirkia* would have lain buried in the Cambrian sea floor waiting for some unsuspecting creature to pass over their tubes. To tread near the entrance of these shafts would have been like stepping on a land mine to most Cambrian animals. The mouths of comb jellies are simply apertures, whereas the mouthparts of priapulids consist of a reversible proboscis, or mouth, and 'lips'. This is obviously more complex and leaves its mark in the fossil record. The proboscis can be withheld inside the head, then extended out into the environment by a process of turning inside out. In this extended position, the lips are revealed at the extremity, along with rows of spines and teeth capable of ensnaring prey. When hooking is complete, the whole proboscis would be inverted back into the head, taking the ambushed prey with it. Most of the Burgess bristle worms also possessed a reversible proboscis, although not one so heavily laden with offensive spines. That is because most Burgess bristle worms fed on organic particles in the sediment or were scavengers on carcasses. But a more complex array of feeding parts leaves greater signs of predation in the fossil record, and such an array belonged to the active predators – those that actively hunted their prey.

A Cambrian arrow worm has been uncovered from the Stephen Formation of the Burgess site. This was buried at a deeper water location than most of the other Burgess fossils, but it was an active swimmer so may also have inhabited shallow-water sites. Interestingly, just as arrow worms are today, then it was also a predator. We know this because it possessed the characteristic arrow worm mouth spines, the tools that grasp prey in mid-water. The prey in this case would have been small and planktonic, but other active predators in Cambrian seas were large, and their grasping tools and mouthparts were fearsome.

The most memorable fossil I examined at the Burgess quarry, protected under its plastic sheet, was a specimen of *Anomalocaris*. One look at its grasping forelimbs and the word 'predator' springs immediately to mind. *Anomalocaris* was widespread at least between 525 and 515 million years ago, when it was the number one predator. At up to 2 metres long it was certainly the largest animal of its time.

Recently, the Japanese Broadcasting Corporation NHK made a full-size model of *Anomalocaris* for a documentary series. The overlapping flaps along the side of the animal were undulated in a wave-like manner, and the model emerged highly manoeuvrable, like cuttlefishes today. It could move forwards, backwards or simply hover in mid-water. So the Burgess and a similar Chinese *Anomalocaris* species were able to actively swim after their prey. An Australian species, on the other hand, was more cumbersome and probably combed the mud for its prey. But all species of *Anomalocaris* benefited from the same type of circular mouth, a collection of hard plates that open and close like the iris of a camera, with a circular array of teeth inside. The aperture itself was rectangular and could not be closed – the teeth did not meet in the middle. Rather the mouth was opened further to admit prey, then the hard plates were pulled together to draw the prey into the mouth. This action would have cracked or even broken the armour of arthropods. Unfortunately, before this reconstruction of a large, fearsome arthropod, *Anomalocaris* and its various parts had to pass through other interpretations. Throughout the history of palaeontology it has been a jellyfish, a sea cucumber, a bristle worm, a sponge and a shrimp. Sometimes it is worth keeping digging.

The five-eyed *Opabinia* is another obvious predator with its moveable, snapping mouthpart. *Opabinia* was just as manoeuvrable in the water as *Anomalocaris* – actually they were probably related. The snapping mouthpart of *Opabinia* probably represents the grasping forelimbs of *Anomalocaris* – twisted 90° with their bases elongated into a tube. Based on the shapes of the bodies and limbs alone, it would appear that the list of active predatory forms represented in the Burgess Shale is long.

Most of the large arthropods of the Burgess Shale were certainly predators, actively hunting their prey in mid-water. Some, like *Odaraia*, do not have large grasping limbs and would have preyed upon shoals of small floating or swimming organisms. Others, like *Sanctacaris* and *Sidneyia*, were armed with a barrage of spines and claws, and would have been formidable predators to most Burgess organisms. But what about the best represented group of arthropods in the Cambrian – the trilobites?

Some Cambrian trilobites had sizeable digestive chambers for the initial processing of food. These, surely, were predators – they needed to

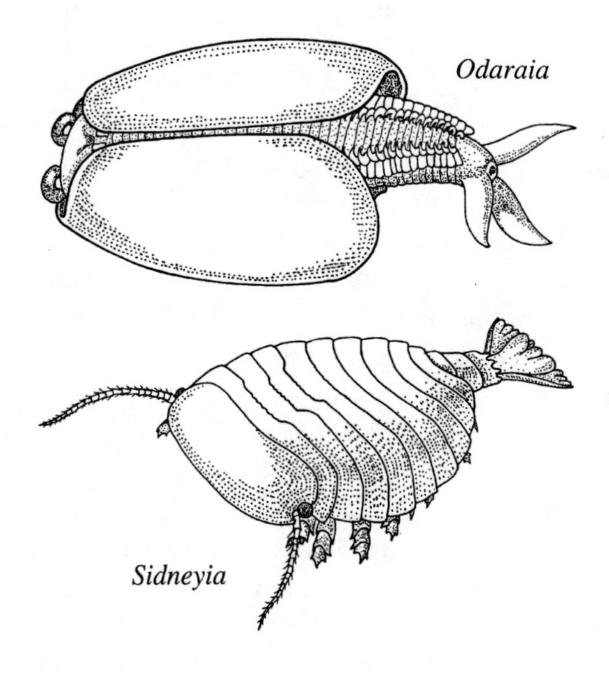

Figure 8.3 *Odaraia* and *Sidneyia* from the Burgess Shale.

store single, large food items at a particular time. Such large digestive chambers are not found in detritus-feeding trilobites, those species that combed the sea floor for particles of organic matter. There were indeed trilobites which employed such feeding methods, and others that were planktivores, filter feeders . . . some even cultured bacteria to provide a meal. Much of the evidence for this comes from the specific shapes of the fossils themselves. For instance, trilobite expert Richard Fortey of the Natural History Museum in London noticed the bulging sides and reduced mouthparts of one trilobite. From this he understood that food was absorbed through the gills along the sides of the body, derived from bacterial colonies living there. Today, crustaceans living at mid-oceanic ridges and hydrothermal vents obtain nutrients from similar bacteria housed in their gill structures. As further supporting evidence, Fortey's trilobite inhabited a similar environment.

Using crustaceans as their modern-day representatives, it would seem that most trilobites were predator–scavengers. That is, they fed on the bodies of other multicelled animals, either living or dead. Their heavily spined, robust limbs could have had no other purpose but to grasp and tear apart whole animals. As will be examined in more detail at the end of this chapter, the majority of *early* trilobites were active predators – they moved rapidly to hunt their prey. Further evidence to support this view is sealed within fossils of the Naraoids.

Naraoids were a sister group of trilobites – they were their closest relatives and bore a physical resemblance. Naraoids too possessed very spiny and formidable-looking limbs, in addition to fang-bearing mouthparts. They probably fed on worms and other soft-bodied creatures. But Naraoids were different from trilobites in one respect – they had relatively soft bodies. The upper surface of their exoskeleton was not calcified like that of a trilobite, but was only organically strengthened. For this reason, to support the massive spiny limbs the upper surface of the body could not have been jointed like that of a trilobite – it would have been too weak. The upper body surface is a point for muscle attachments and is comparable to the supporting walls of a house. So the predatory limbs of Naraoids came at quite a cost to their bodies.

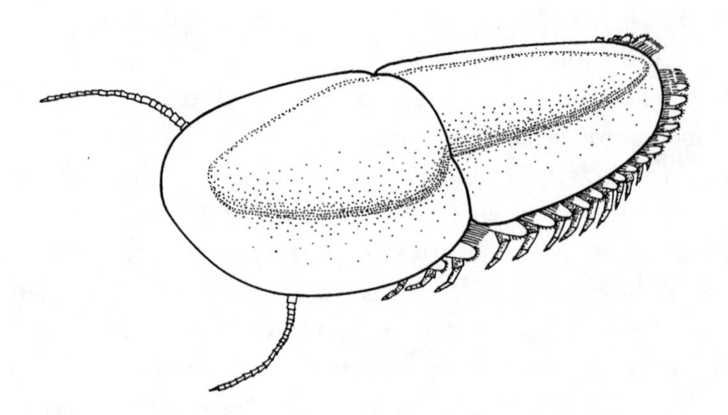

Figure 8.4 *Naraoia*, a Naraoid from the Burgess Shale.

Naraoids evolved from the ancestor of the trilobites before the exoskeleton became strengthened with calcium carbonate, or calcite. Naraoids are often found distorted in their burial grounds as a result of their weak exoskeletons, and they also missed out on eyes. The important evolutionary history of Naraoids and trilobites will be considered further in this chapter, but first we should examine the remaining evidence for predation in the Cambrian. So far we have looked at eyes, feeding apparatus and digestive systems. Now we should look for teeth marks in the prey.

The display of Burgess Shale fossils at the information centre in the Canadian town of Field contains an interesting trilobite. Although most fossils here are remarkably complete and favourably orientated in the rock, and include some of the best examples of the Burgess species, a specimen of the trilobite *Olenoides* is particularly noteworthy. A large part of its body is missing, but the regularity of the semicircular omission suggests this was not an artefact of preservation. It could be only one thing – a bite mark. A large Cambrian predator bit this trilobite – it was Cambrian prey.

Many other Cambrian trilobites have been found with scars, signs of attack by a predator sustained while still alive. These wounds proved not to be fatal because of the animals' ability to heal. This is an interesting concept in itself. Cambrian trilobites were well prepared for attack not only in their protective armour but also in their ability quickly to seal the newly exposed body sections – they could form calluses. Human skin is thin and can be easily cut. For this reason, our blood has the ability to coagulate and seal up broken blood vessels, preventing blood loss and infection. Arthropod exoskeletons, on the other hand, are tough and designed to withstand the rigours of their hosts' lifestyles . . . except when they are heavily attacked. The self-healing of Cambrian trilobites indicates they were so prone to attack that predation had certainly been a selection pressure during their evolution. Today animals can be found with hard shells that have functions other than to protect them against predators, such as providing support for tissues. But not only had Cambrian trilobites evolved armour, they had also evolved a self-healing mechanism to function in the event of attack by predators. Their hard shells had a role in protection against predators from the beginning.

There have been so many Cambrian trilobites found with bite marks that a theory of 'handedness' has been suggested. In a large sample of trilobites, seventy-seven specimens had sustained injuries of unknown origin, perhaps caused by accidents during moulting or mating, but eighty-one specimens revealed injuries caused by predatory attacks. Researchers at Ohio State University found that 70 per cent of all scars left by predators were on the right side of the trilobites. It is thought that trilobites, their predators, or more likely both, tended to favour one side. Trilobites probably veered to one side in an attempt to evade an attacker. Also, predators probably tended to attack from the same side. Such asymmetrical behaviour is commonly seen today – a horse tends to turn its head to the left and 90 per cent of humans are right handed. But of most relevance to this chapter is the shape of the trilobite scars, whether on the left or the right side of the body. Many were W-shaped, conspicuously matching the size and shape of the iris-forming, triangular mouth plates of *Anomalocaris*.

With the exception of the trilobite-like *Naraoia*, all the Burgess arthropods were protected within armour. They possessed head shields that were sometimes further protected by solid bumps or spines. Many trilobites possessed large spines – the defensive role of these becomes obvious when trilobites are considered in their curled-up posture. Here the animal is transformed into a hard ball with projecting spines. Furthermore, the spines are sometimes quite elaborate, with serrations and spikes.

Long, sharp spines can be found on many hard exoskeletons, but also protecting softer, more fragile bodies like that of the Burgess lace 'crab', *Marrella*. In fact the use of armoured spines to protect soft bodies was employed by animals from a range of diverse phyla. There was the velvet worm *Hallucigenia* with its soft body protected by long, upwardly projecting spines. The bristle worms are more classical cases of this phenomenon; *Canadia*, for instance, wore a coat of spines projecting upwards and sideways. It is thought that bristle worms and their relatives in freshwater today independently evolved spines for defence purposes. Such convergence suggests that this means of protection is a good one. The epitome of protection within the Burgess bristle worms, however, was to be found in *Wiwaxia*, an oval-shaped

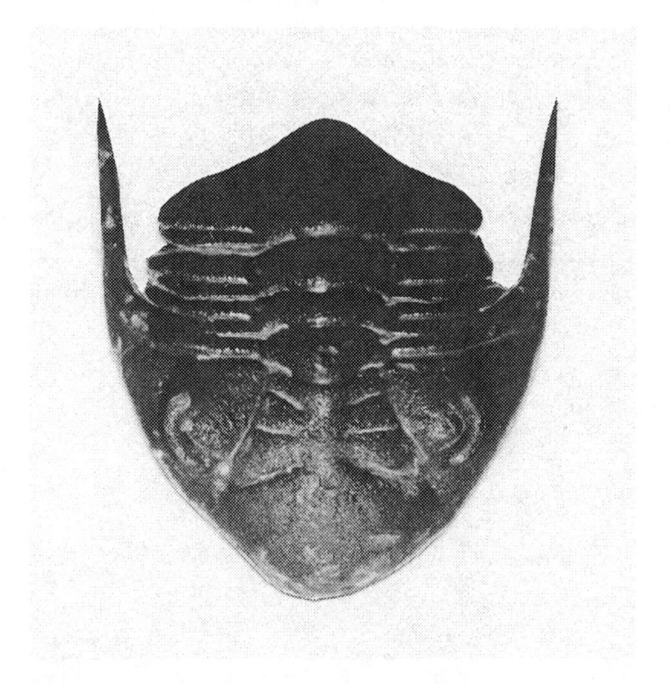

Figure 8.5 Photograph of a trilobite when rolled up – 'head' spines can be seen projecting from the body. When the trilobite is flat, as we usually view trilobites, these spines lie flush with the body.

animal not only completely covered by overlapping shields but also with long swords projecting outwards that made even *Hallucigenia* look like easy pickings. Halkieriids are possible ancestors of, and bore a close resemblance to, *Wiwaxia* – they possessed a similar means of protection in their chain-mail coat of shields.

The sponges *Choia*, *Halichondrites*, *Pirania* and *Wapkia* of the Burgess Shale contained spicules that not only provided a supporting lattice, but also projected into the environment as deadly blades. Burgess priapulid worms had spines in the region of their mouths for feeding, but also on other parts of their bodies, where they took their most fearsome forms. Like most lamp shells, the Burgess hyolith *Haplophrentis* completely closed shop by surrounding its entire body with exceptionally hard armour. The Burgess echinoderms, relatives of starfish today, similarly revealed no soft parts to a passing predator.

And finally we can consider *Micromitra*, where a shell as hard as a mussel's was obviously not enough to escape predation – it further evolved long spines around its edge.

An earlier lamp shell, *Mickwitzia*, may have taken protection a stage further. *Mickwitzia* possibly employed chemical defences – it squirted toxins through holes in its shell. The evidence for this derives from the other shelly fossils found with *Mickwitzia*, which all exhibited boreholes made by predators. *Mickwitzia*, on the other hand, was always borehole free. To conclude, all of this evidence can mean only one thing – that animals possessed protection against predators in the Cambrian.

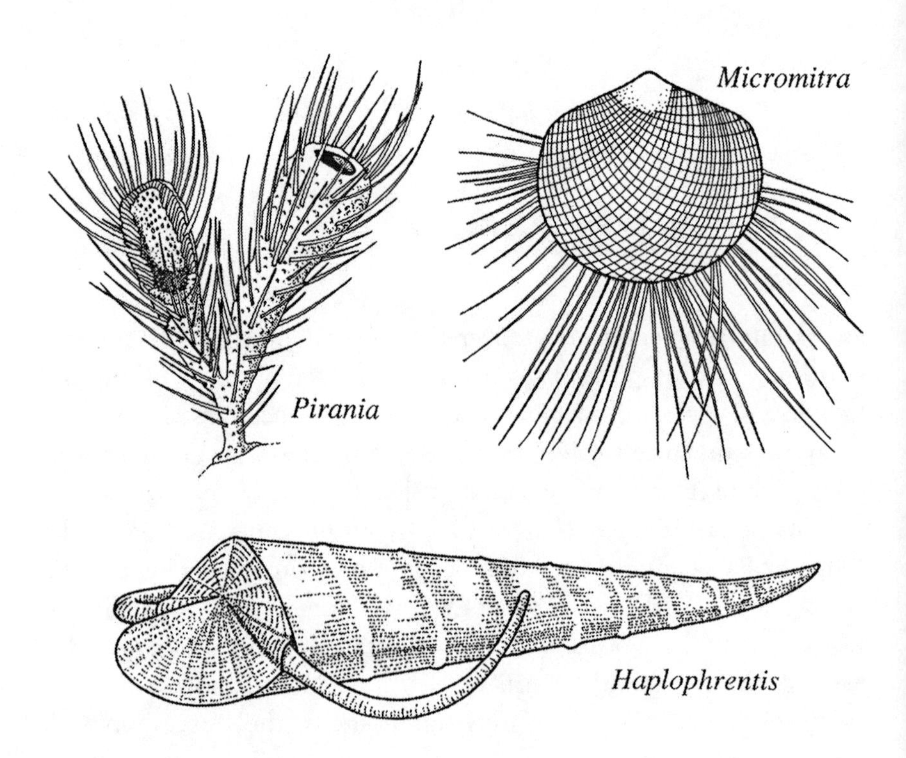

Figure 8.6 *Pirania, Micromitra* and *Haplophrentis* from the Burgess Shale.

The hard parts described so far all evolved at one point in time. This evolution was the Cambrian explosion – all animal phyla suddenly evolved their hard parts simultaneously between 543 and 538 million years ago. As mentioned already, hard parts can have functions other than to provide protection against predators, but it would appear extremely coincidental for all phyla to evolve hard parts at precisely the same time to provide strength or as a barrier against osmotic stress. Multicelled animals from different phyla had been around, in soft-bodied form, for 100 million years or so beforehand. And as established in Chapter 1, physical environmental conditions that could have demanded hard parts were not the cause of the Cambrian explosion. Now it becomes important to chart the original appearance of predators, particularly the highly active forms. This will be investigated as soon as all the clues from the Cambrian have been gathered.

The fact that all Burgess arthropods possessed protective spines, or some form of protection against attack, means they were not only predators, they were also prey. With the exception of the top predator *Anomalocaris* (which lacks protective spines) it is not surprising we could not deduce whether most Burgess animals with eyes were predators or prey based on their optics. In fact the ambiguity in the optical data supports the idea that most Cambrian animals in the open water were looking out for both prey and predators. With *Anomalocaris* a common menace in the Cambrian, the first rule was to stay alive, which meant keeping a lookout for the big-eyed giant. Cambrian eyes must have been adapted for scanning the complete environment, and any modification to this must have been slight due to the wrath of *Anomalocaris* and other highly mobile predators. That, as it happens, is exactly what we found in the Burgess eyes – adaptation for 360° of vision, with minor directional qualities.

As for who ate who exactly, we can assume that the larger swimming forms preyed upon both the smaller swimming forms and the soft-bodied bottom-dwellers. But in addition to the telltale scars of *Anomalocaris*, there are other signs of predation in action where we can solve this problem more precisely. Fragments of the Burgess hyolith *Haplophrentis* have been discovered in the guts of thirty individuals of *Ottoia*, a priapulid worm, and also in the gut of the large arthropod

Sidneyia. In that same gut of *Sidneyia* have been found seed-shrimps and trilobites – *Sidneyia* could feed on hard-shelled animals. And a closer inspection of the gut of one *Ottoia* revealed part of another *Ottoia* – so this priapulid worm was a cannibal.

One fossil I picked up from the display table at the Burgess quarry was also interesting from this respect. This was the shrimp-like crustacean *Canadaspis* ... and the tiny trilobite *Ptychagnostus*. The trilobite lay within the rounded head shield of *Canadaspis* and was probably its dinner. Other small trilobites have been found within the head shields of other Burgess arthropods, and it is possible that they were parasites. Since *Ptychagnostus* is found commonly in isolation, and probably lived in mid-water, maybe it was both parasite and prey.

It is worth a closer look at this situation from another perspective. *Canadaspis* has eyes whereas *Ptychagnostus* does not. *Canadaspis* and *Ptychagnostus* evolved to gamble on different aspects of The Laws of Life. *Canadaspis* placed its chips on 'eating', *Ptychagnostus* on breeding. *Ptychagnostus* was extremely common in the Cambrian, whereas *Canadaspis*, and all other large Cambrian predators, was far less numerous. *Ptychagnostus* as a species was obviously prepared for predation – its survival was dependent on numbers. In other words, *Ptychagnostus* must have evolved a successful breeding strategy, so there were more individuals living in the water than could be consumed by predators – the strategy taken by krill in response to baleen whales. But this bottomless-pit-of-food scenario was not so accommodating to *Canadaspis* and its fellow predators. There remained the simple matter of 'search and destroy'.

The open water is three-dimensional. One is less likely to bump into an animal in the open water than on the sea floor, which is a two-dimensional environment. The ocean is vast, and *Ptychagnostus* mobile. To catch *Ptychagnostus* one must have the ability to find it and swim after it. Thanks to the Cambrian explosion, the strong, skeletonised limbs with internal muscles achieved the mobility required. And like dragonflies in the air today, the eyes of *Canadaspis* gave it the means to find. Here we are building a picture of how life functioned in the Cambrian, and the rules are similar to those today. In Chapter 3 it was revealed that eyes and armour are not directly related in that species with eyes do not always have armour, and vice versa. However, there may be a link in behaviour . . . and evolution.

Chapter 1 introduced an old idea for the cause of the Cambrian explosion, one reworked recently by Mark McMenamin and Dianna Schulte McMenamin – that a food web was developed at the beginning of the Cambrian, where every species had its own predators and food. But can entire food webs simply spring up out of nowhere? Or is there a factor that triggers a chain reaction, ending in the formation of a full-blown food web? The range of animals living just after the Cambrian explosion, with their diversity of shapes and sizes, suggests a mature food web was in place at that time. But did this food web mature quickly, or instantaneously as the Cambrian enigma demands? Or did it assemble gradually, beginning in the Precambrian? These questions indicate that our Cambrian jigsaw puzzle is nearly complete.

The McMenamins resurrected a century-old idea that animals developed shells as shields against predators, a fact we too have established. In this chapter, predators are emerging as hugely important factors in the way life works today and how it did in the past. But when did the remaining feeding modes within food webs first appear? This question may be irrelevant to our overall quest if food webs, and consequently The Laws of Life, were established *before* the Cambrian. It would seem appropriate for this chapter to end in the manner of the previous one, in a search for the *beginning* of predators on Earth.

In the original line of fire

Journeying beyond the Cambrian explosion and into the 'relative unknown' that is the Precambrian, the first port of call is the age of Ediacara. The Ediacaran suite of life forms is best represented by the original finds from South Australia, around 565 million years old, although the same organisms existed right up until the Cambrian explosion itself, when they disappeared without trace. But while they existed, they did exhibit a variety of lifestyles. We know this from the shapes of the life forms themselves and from their trace fossils – footprints and their equivalents.

The Cambrian was once described as a peaceful time, but we have now established this is not true. In fact we know now that predators existed even before. In the Precambrian there were jellyfish pulsating through mid-water, and relatives of the Portuguese man-of-war floating on the surface. Any creature that accidentally encountered the stinging tentacles of these animals would have instantly become their prey. On the sea floor were anemone-like creatures with their stinging tentacles waving expectedly upwards. And then there was Precambrian prey. In some cases the Ediacaran predators would have preyed upon each other. But there were also flat, worm-shaped animals that probably undulated their bodies to propel them through the water – and sometimes into the lions' den. Occasionally they would have propelled themselves into the nets of stinging tentacles, cast hopefully into the water.

Although the word 'hopefully' infers personality in these primitive forms, it *is* appropriate in that Precambrian predation was a comparatively random process. There was no *Anomalocaris* with its advanced detection system and search-and-destroy capabilities. All that patrolled the Precambrian water were the stinging nets of jellies. But in the jelly's favour, the prey could not sense them coming either.

This last statement may not be strictly true. Although too small to be recorded as fossils, Ediacaran organisms surely possessed sense organs of some kind. They may have sensed vibrations in the water, based on movements of tiny hairs on their skin, which could signal the advance of a barrage of stinging cells. Indeed, the probable relatives of some Ediacarans today are endowed with hairs of this type. But detection in the Precambrian would have been possible only in close encounters. And selection for a more advanced sensing system would have been minimised by the generally slow speed of the advancing predators. This was a kitten-and-mouse game, in comparison with the cat-and-mouse Cambrian.

On the sea floor the threat of predation was no more severe . . . but did exist all the same. A worm-like animal called *Claudina* lived in the sediment just prior to the Cambrian, about 550 million years ago. It is known from precisely 524 fossils from Shaanxi province in China. The fossils are not of the animal itself, but of its tube – this is the first animal known to possess hard parts. It appeared to have jumped the gun before the start of the Cambrian, at the same time demonstrating that environmental conditions were not completely restrictive for making hard parts before the great explosion.

Fourteen tubes of *Claudina* revealed boreholes – holes made by a predator on the sea floor in a successful attempt to consume the soft animal within. Stefan Bengtson of Uppsala University and Yue Zhao of the Chinese Academy of Geological Sciences, who found these fossils, believe the predator to have been a mollusc, possibly a relative of snails today. But in the Precambrian, molluscs, like most other animal phyla, looked like 'worms', or rather had completely soft bodies. There was not even a hint that one day its descendants would carry around huge shells.

The holes in the tubes of *Claudina* provide the first definitive evidence of predation on Earth. And it seems that what can be best described as 'inactive predation' was common in the Precambrian. Although, based on the lack of armour worn in the Precambrian, this type of predation obviously did not present a strong selection pressure for counter-predatory measures. It did not provide the stimulus for hard, protective parts.

In particular, there was one interesting soft-bodied animal that roamed the Precambrian sea floor. In 1984, petroleum companies were exploring parts of southern Morocco and eastern Siberia. They drilled vertically into the ground and removed cores – long, thin cylinders of rock that revealed the layers of sediment built up over 600 million years, while these areas were underwater. As expected, rocks that formed just before the Cambrian showed signs of stromatolites. But there were further, unusual layers just above the stromatolites. At the time they were termed 'thrombolites' and were assumed to be the result of grazing by soft-bodied arthropods, including 'proto-trilobites'. Indeed, the first signs of trilobites, the first hard parts of any type, were found some tens of metres above the lowest thrombolites. Along with trace fossils of soft-bodied arthropods, this is an important clue in piecing together a picture of the ancestors of arthropods. But most enticing is the term 'proto-trilobite'. Did trilobites as such exist without their armour before the Cambrian explosion? In 1991 this

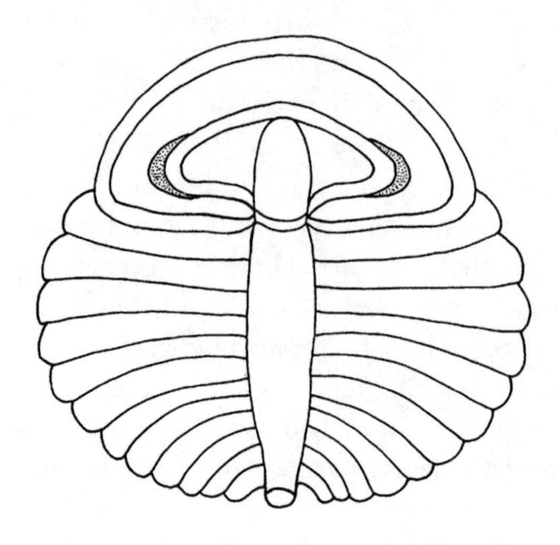

Figure 8.7 A soft-bodied 'trilobite' from the Precambrian (about 565 million years old). Shaded regions in the head could be the precursor to compound eyes.

question was answered. Incredibly, a new expedition to the original Ediacaran site, the Ediacaran Hills in South Australia, yielded a soft-bodied trilobite.

First, raking trace fossils were found from an animal predicted to be 4 centimetres long with twelve pairs of long, thin legs. Then came the real breakthrough. Several specimens of the bodies themselves were discovered – the originators of the trails. The bodies would have been soft compared with those of Cambrian trilobites. From above they were round, but showed clearly a semicircular head with a well-marked border, a thorax with thirteen large segments and eight smaller segments, and a tiny, oval tail. Some specimens were distorted, indicating a degree of elasticity in the skin like that of the Cambrian Naraoids. But the general body architecture matched that of Cambrian trilobites – except for an elastic skin in place of the hard exoskeleton. Also of interest in the proto-trilobites were curved, shallow ridges on the head, in the region that eyes were housed in Cambrian trilobites. But eyes themselves, like grasping limbs and spiny mouthparts, were absent in the Precambrian forms.

The proto-trilobites of the Precambrian were grazers, feeding on algal mats and probably dead animal matter lying on the sea floor. It seems the voracious predators that emerged with the Cambrian had rather peaceful beginnings. If anything, the proto-trilobites would have been prey themselves – the tables may really have turned at the Cambrian border. In general the Precambrian was rather an experimental stage for predation, occupied mainly by peace-loving vegetarians that were willing enough to accept any occasional animal matter they stumbled upon. For they were developing a taste for meat.

Was this shift in emphasis towards predation a gradual phenomenon? It seems not. Carnivores really made the headlines 543 million years ago. Suddenly, predation was not only a major option within the food web; it encompassed a new form. If the Precambrian predators were considered passive, the second wave of predators that swept through the early Cambrian seas were undeniably active.

The end of this chapter will duplicate the end of the previous chapter, where we learnt that the first animal with eyes was a trilobite – the first trilobite. The first true trilobite was also a predator. *Fallotaspis*, *Neocobboldia* and *Shizhudiscus*, all trilobites with eyes, were also icons of the beginning of the Cambrian, around the time the Cambrian explosion began. Their limb shapes indicate that these trilobites were predators; their spiny shields affirm that they were also prey. They probably attacked each other – the archetypal attacks on Earth, since their bodies were armoured in only rudimentary form. Their skins had become less soft than those of the Precambrian proto-trilobites, but they were still not fully hardened, as were exoskeletons of trilobites that appeared a few million years later. They were, however, highly active animals. They could swim rapidly, they could manoeuvre in mid-water . . . and they were predators with spiny, robust limbs. They were bad news for Precambrian-style, soft-bodied forms everywhere. Life was about to be stirred up.

So the beginning of the Cambrian was also the beginning of *active* predation. This is a simple concept that warrants little additional discussion. But there is one detail we should consider further – we must distinguish in our minds the difference between the Cambrian explosion and the cause of the Cambrian explosion. The signs of predation we are using to denote the beginning of active predation are the spiny tools and swimming limbs of trilobites – or hard parts. But the acquisition of hard parts *was* the Cambrian explosion. This chapter will close with a question: 'Did a few species of predatory trilobites evolve from proto-trilobites and kick-start a chain reaction?' That chain reaction, of course, was the acquisition of hard parts and other external characteristics in all animal phyla – the Cambrian explosion. Did all hell break loose simply on the appearance of a few armoured forms, or did something else happen that sparked the evolution of hard parts simultaneously everywhere? It is finally time to put two and two together.

9

The Solution

> From a perception of only three senses or three elements, none could deduce a fourth or fifth
>
> WILLIAM BLAKE, *There is No Natural Religion* (1788)

The sun emits a continuous array of electromagnetic waves – radiation ranging from cosmic and gamma rays, with wavelengths smaller than an atom, to radio waves with wavelengths of over a thousand metres. Visible light waves lie within this spectrum, at the peak in the sun's energy emission. They include only a narrow range of wavelengths. When light waves fall upon an object, they can be deflected and relay news of that object into the environment. If the deflected waves meet our eyes, they can be focused on to a retina, and we can interpret the news. One item of news that helps us to 'see' is the direction from which these waves last came. This we can determine simply. With two eyes, we can also judge the distance of the object deflecting the waves. But a third trick of the eye is to convert light waves varying slightly in wavelength into different colours. So for an animal without eyes there is no such thing as colour in its environment.

This is difficult to comprehend. But just think: all those wonderful colours we see around us, wherever we are, do not actually exist. In the environment there is *no* colour, only objects that happen to deflect different types of electromagnetic radiation. Roses are not beaming out reds, nor do leaves generate greens. Perhaps the one chance we have of dealing with this truth lies with ultraviolet.

To birds and insects there is even more happening in the environment, even more colour. Their palette also contains ultraviolet – they are communicating with private wavelengths, oblivious to us. But birds and insects could not comprehend that some other animals cannot detect ultraviolet light. So in turn we should remember that not all animals see images nor understand what *we* mean by colour. That's not to say that light and colour are not a big part of the lives of *all* animals. The word 'colour' can be found in the dictionaries of all animals living where light exists. Although not all are conscious of the fact, light is a major selection pressure acting on *everyone* . . . or at least it is today.

Plants are governed by very different rules to those of animals, yet even many plant colours are adaptations to animal vision. Leaves generally have to be green because their component chlorophyll deflects the wavelengths that we interpret as green (those wavelengths not used for photosynthesis) – this is incidental colour. But many plants produce flowers that display a vast array of colours to attract pollinating insects, and also colourful fruits to attract seed-dispersing mammals and birds. In fact, animals with eyes may even provide the main selection pressure in the evolution of some plant groups. For instance, the flowers of the *Ophrys* orchids have evolved to mimic females of different species of *Campsoscolia* wasps in terms of colour and shape. This mimicry is so effective that the male *Campsoscolia* wasps are deceived and attempt to mate with the flowers, but succeed only in transporting pollen.

In his book *The Universe of Light*, published in 1933, Sir William Bragg introduced the concept that 'Light brings us the news of the universe.' Light is not the only messenger, or stimuli, on Earth – there are other conveyors of the news of the universe, most notably sound and chemicals – so it needs to be put into perspective. It may be useful to make analogies between the natural stimuli and the different forms of media that supply our political news (excluding the Internet).

We can receive our daily political news from television, radio and newspapers. The producers of the news in these three different formats operate very differently. In terms of history, newspapers were the first to appear. Reporters roved around newsworthy scenes, and brought home their stories on paper. Their job became easier with the introduction of the telegram machine and telephone. In fact their job changed a little following these innovations – reporters 'evolved' in response to their changing environment.

The introduction of radio saw further changes to the reporters' technique, but previous technology could still be used, albeit in a new way – telephone messages could be broadcast directly. But now *all* the print could be read out into a microphone, and the news producers' job had changed, or 'evolved', once more. Small improvements in technology translated to equivocal developments in the news service. As technology advanced, the news producers responded to adapt to their new environment. If they did not adapt, they would have been overtaken by rival companies. They would have been forced to target an unenviable minor audience, or elbowed into a remote and limited niche. 'Micro-evolution' was taking place in the world of news broadcasting.

Then came a momentous change – television was invented. The news producers' job had to evolve once more . . . but this time dramatically. New equipment was needed, along with new people with the skills to operate it. Old-style reporters were replaced with non-camera-shy reporters, who also met new demands on visual appearances. New buildings and vehicles were required. Basically the whole news scene changed – a different type of worker was required at every position. There had been a 'macro-evolutionary' event in the conveyance of news that turned the trade upside-down. The gradual changes that had been taking place in the other types of media now would seem trivial in comparison.

The introduction of television happened almost overnight. Some significant changes happened subsequently, such as the conversion from black and white to colour, and the introduction of satellites, but eyes were already focused on the television at news time, and *that* event was the *really* big one (again, the Internet excepted).

The introduction of television to the field of news broadcasting would have had even greater impact if everyone on Earth had possessed a television set. The effect would have been similar to everyone suddenly evolving eyes overnight. That's an interesting thought, and a concept that also applies to this book, particularly when considered with the previous comment that light is a stimulus for all animals on Earth . . . *at least today*.

The link between the power of light and the behaviour and evolution of all animals today is the eye. Eyes make light a stimulus for everyone – even individuals without eyes. Today the importance of eyes to animals living in sunlit environments is often considerable, as is evident from the size of most animals' eyes. Dragonflies have big heads, with eyes occupying three-quarters of the area, and some seed-shrimps have eyes which monopolise a third of their body volume. And a large proportion of the brain of eyed animals is always devoted to vision.

When the first eye was traced, it emerged that it belonged to the first trilobite, or 'last' proto-trilobite, and appeared at the very beginning of the Cambrian explosion. There is a link here which suggests that eyes may have been the 'television' of evolution, and it is one that cannot be ignored.

Should we consider predation too?

The last chapter introduced a new variable to the equation – feeding. It also threw a real spanner in the works of the neat theory that was forming. It shed further light on that first trilobite to evolve at the beginning of the Cambrian. This was the first animal with eyes, but it was also both predator and prey. We learnt that predation was evolving gradually during the Precambrian. But the first trilobite was the first highly active predator. This is different. It means that another factor – active predation – can also be associated with the beginning of the Cambrian explosion.

So where do we stand now? Are two possible causes of the Cambrian explosion developing in this book? First we should examine these possibilities further, and perhaps try to integrate the evidence.

Consider the military expression 'search and destroy' used in Chapter 8. The word 'search' precedes 'destroy', and that is exactly the order of action in the process of active predation. Before destroying, one must search, identify and capture. Active predators would be useless without eyes or a comparable detector for another sense. At the beginning of the Cambrian, animals began madly chasing and eating each other. A prerequisite for this behaviour is an appropriate search capability – the attributes of speed, agility and grasping hooks would be redundant without a knowledge of where the prey is. And indeed, at the beginning of the Cambrian predators first set their sights on prey. Setting sights, as in getting a victim within the telescopic sights of a rifle, is an appropriate term because the early Cambrian killers did place their victims within sights – their eyes. It seems that Chapters 7 and 8 are beginning to overlap. Now the long spines extending from the bodies of many early Cambrian trilobites can be interpreted.

Armaments are ornaments

Emphasis has been on the great importance of light as a stimulus to animal behaviour today. In fact all of the terrestrial animals (excluding domestic species) we are familiar with are wonderfully adapted to light not only in terms of their colour, but also their behaviour and sometimes shape. Colour is the logical animal adaptation to light, and the external colour of an animal living in an environment with light is usually an evolutionary response to that light. For instance, it is argued that in spiders the production of colour is chemically costly and is principally maintained by the action of sight-hunting predators. Shape, on the other hand, is largely governed by chemical processes, movement, reproduction, feeding mechanisms and other behaviours. But for some of these activities light may also be a major consideration. Here behaviour becomes important. A stonefish not only has to be coloured like a stone, but must also have a similar shape and behave similarly, spending long periods stationary. Also praying mantids possess the colours and shapes of the plant parts on which they live, whether they be green leaves or pink petals. Then there are the stick and leaf insects, which are related to the praying mantids but are the hunted rather than the hunters. Stick and leaf insects possess the light adaptation characters of colour and shape, but unlike stonefishes and praying mantids they must move to find food. And to complete their adaptation to vision, they walk with the quivering movement of leaves or petals in the wind.

Once again we are led to consider eyes, albeit those belonging to animals other than those in question. The above mentioned colour, shape and behavioural characteristics are not directly an adaptation to sunlight, but rather adaptations to the presence of animals with eyes. But in particular it is the eyes of either predators/enemies or prey. There is a potent relationship between eyes and predators, or between the visual appearance of animals and staying alive. Staying alive, according to The Laws of Life, can mean eating and/or avoiding predation.

We can now understand why camouflage is common among animals today. Many insects are green so as to be camouflaged against leaves. Although green is generally a difficult colour to achieve, pea aphids *are* green where their predators, ladybirds, abound. Ladybirds hunt mainly using vision, and so camouflage is a good strategy for their prey. But when ladybirds are scarce, the pea aphids stop producing the energy-expensive green pigments and turn a less costly red. Similarly, guppies change their visual appearances in response to predators with eyes. Populations of this fish, found in Trinidad and South America, vary markedly from each other and so have become classic animals for the study of evolution in action. A population can transform in terms of colour and anti-predator behaviour within a few years, or ten generations, of a change in predator pressure. Of course mating is another important behavioural and evolutionary consideration, leading to sexual selection. Sexual selection acts in unison with predator-driven evolution, or natural selection. When the threat of predation is relaxed, bright mating colours will evolve in guppies via sexual selection. But all of this evolution is driven by vision, whether the vision of other guppies or of their predators.

Mating leads to well-known exceptions to the rule of camouflage, particularly in birds, where vision is usually the primary sense. Consider the peacock, where Newton's analysis of colour applies only to the spectacular males with their imposing tail feathers, not to the dull-brown, short-tailed females. Yet both sexes of peacock share the same feeding strategy. A key element here, however, is the relatively modest threat of predation, and this is a luxury afforded to most birds. Flight in vertebrates has generally provided an evolutionary 'time out' from the camouflage constraints imposed upon most animal species on land and water. So many birds are free to display colours suitable for another important behavioural process – courtship. And as could be predicted from this philosophy, birds have evolved some of the most sophisticated, visually oriented courtship displays. They can stand somewhat clear of the cat-and-mouse world sculpted by the presence of predators with eyes.

Back on the ground or in the water, The Laws of Life are far stricter. There are no magical hiding places or extra dimensions into which animals coloured with maladapted hues could instantly vanish. But at the same time, light paves the way for increased adaptive radiation here. Cases discussed in this book have included those of the East African Rift lake cichlid fishes and Caribbean *Anolis* lizards. Adaptive radiation involves movement into different available niches. Light generally creates more available niches – shade and bright light, and different coloured backgrounds, for instance. Hence sunlit environments support a greater diversity of animal life than do cave environments.

Put together all of the considerations listed and we have a world where light shapes most ecosystems. Consider the marine environment. One can choose to live in different light regimes. There is the sea floor to burrow into, or crevices in rocks and corals. Similarly, sponges provide suitable hiding places, and the stinging tentacles of anemones or Portuguese men-of-war can be another safe option (if one is immune to their toxins). Then one can be brave and shun the protection afforded by external sources, but living out in the open potentially places one in the line of fire. So a survival strategy must be evolved to reduce the risk of predation. One may be camouflaged or transparent. Then there is the conspicuous option – don warning colours or protective armour. Or one can be fast and on the ball, capable of spotting and outrunning any predator. Alternatively one can concede defeat to predators, and choose an unusually successful breeding strategy at the expense of reducing individual chances of survival. At least this way one's species may survive (although this would not work if employed by *all* prey species, since 'space' for this niche is limited). But either way, a good strategy to counter those predators with eyes is essential.

Although this is not strictly the language of an evolutionary biologist, it does sum up the idea of selective pressures that act on evolution. And all of this comes about because predators exist with eyes. Without eyes, light would not be a major stimulus to animals.

At this point I feel like a university lecturer who has just finished teaching a foundation course – weary but relieved. Not a single educational stone has been left unturned in the bid to reveal the facts and figures needed to progress to a new stage in learning. There is a certain amount of relief because this is where things become interesting and exciting. We are now equipped to tackle evolution's grandest event of all. We can now go back 543 million years, to the beginning of the Cambrian.

The 'Light Switch' theory

Consider dividing geological time into two parts – pre-vision and post-vision. The boundary separating these parts stands at 543 million years ago. Considering vision as the most powerful stimulus on Earth, the way the world functions today is the same way it functioned ten million years ago, 100 million years ago and 537 million years ago, after the Cambrian explosion. Similarly, the world was without vision 544 million years ago just as it was 600 million years ago. In the interval of life's history of these two parts, a light switch was turned on. For the second half it remained on, although during the first half it was always off.

We know that vision places major restrictions on the external forms of animals today, but before the Cambrian it could not have played such a role because eyes did not exist. Consequently light did not exist as a major stimulus in the behavioural system of animals. By vision I mean the ability to produce visual images, which can be achieved only by animals with *eyes*. Light is used to determine the direction of sunlight in numerous forms of simple animals. Testament to this are the algae found in the snow at the Burgess quarries in Canada, with their red eyespots but lack of vision. But these have nothing to do with vision. Indeed, some plants even possess simple light perceptors that regulate the shift from vegetative growth to floral development. But this form of light detection is not vision. Vision is the capacity to perceive and classify objects using light, or seeing.

The Precambrian was a time where only soft-bodied representatives of the multicelled animal phyla existed. On the following pages is a snapshot of life in a Precambrian environment as pictured by the most advanced form of light perceptors of the time.

Effectively light as a major stimulus is, or rather visual appearances are, removed from the Precambrian environment because the animals of that time did not possess eyes. Presumably Precambrian animals possessed chemical, sound and/or touch receptors. They may also have possessed simple light perceptors, like the algae in the Canadian snow, but nothing that could form an image. Light could be considered a very minor selection pressure in the Precambrian. It could not have had a direct effect on the evolution of multicelled animals (it could have had an indirect effect in that animals which fed on photosynthetic algae would have been restricted to sunlit zones).

Competition and predation would not have been major selective pressures in the Precambrian, but they were taking a foothold. The Ediacaran animals of the Precambrian were gradually developing brains. They were developing ways to pick up environmental cues, or news items, and process that information. They were also evolving the ability to chew, and were gradually developing a rudimentary form of rigidness in their limbs. Precambrian trace fossils or footprints suggest that legs could support bodies off the ground. But as in dark caves today, evolution in general would have been slow in the Precambrian, and may well have continued at a gradual pace had it not been for a single but monumental event. This was an event that, in terms of body parts, would have seemed like any other evolutionary innovation, of which there have been many. But this event was different – it changed the world forever on a scale not since witnessed. At the end of the Precambrian, while most phyla were evolving gradually, a serious transformation was taking place in the soft-bodied trilobites. A light sensitive patch was becoming more sophisticated. It was dividing into

Figure 9.1 (overleaf) This is how all Precambrian animals would have pictured their neighbours using light as a stimulus.

separate units. The nerves servicing each unit were becoming more numerous, and so too were the brain cells they serviced. These nerve and brain cells were either multiplying or being borrowed from the wiring and processing system of another sense. Then the outer covering of each unit began to swell and take on focusing properties. One day all this reached a crescendo – a compound eye had formed.

Let there be images! A new interpretation of a sense had entered the animal world . . . but this was no ordinary sense. What was to become the most powerful sense of all was unleashed with the birth of one individual proto-trilobite (during its transition to a trilobite) – the first to entertain an eye. For the first time in the history of the Earth an animal had opened its eyes. And when it did, everything on the sea floor and in the water column was effectively lit up for the first time. Every worm crawling over every sponge, and every jellyfish floating through the water, was in an instant revealed as an image. The lights on Earth were switched on, and they put an end to the gradualness of evolution that had characterised the Precambrian.

Simply put, the visual appearance of animals suddenly became important with the introduction of eyes. But it took just a single pair of eyes – the first eyes – to introduce vision as a stimulus to the world around them, including all its inhabitants. Now if we add vision to the Precambrian scene depicted in Figure 9.1, the animal inhabitants appear as shown on pages 274–5.

The most powerful sense of all had been launched on Earth. Suddenly, and for the first time, an animal could detect everything in its environment. And it could detect it with pinpoint accuracy.

The difference between the previous two pictures, or light perception in the Precambrian and Cambrian, is comparable to that experienced when we close then open our eyes. With our eyes closed we can determine the direction of sunlight but we cannot, for example, find and identify a friend. So, using light, some Precambrian animals could have known which way was up in mid-water, but they could not have found a friend or foe. Nevertheless, in their favour, a potential predator could not have found them either. So there were no strong selective pressures for Precambrian animals to become adapted to light, even though light was to become the most powerful stimulus of all. In fact it became the most powerful stimulus of all almost overnight (in geological terms), with the evolution of the first eye at the beginning of the Cambrian.

With our eyes open, suddenly we see the world very differently. We can see food from some distance, although we can only smell it if it produces a smell, hear it if it produces a sound and touch it if we are very close. So in the Precambrian, not releasing certain chemicals or producing sounds was enough to avoid a potential predator, unless it was bumped in to. But in the Cambrian life was lit up. The light switch was turned on, for the first and only time – and it has been on ever since. With our eyes open we see the size, shape and colour of animals, but we also see their behaviour – we can judge how fast they can move and whether we can catch them. All of these animal attributes suddenly mattered at the beginning of the Cambrian, when the first active predators with eyes were introduced on Earth. At that very point all animals had to become adapted to light, or vision. Near the end of the Precambrian, selective pressures had been acting on proto-trilobites to evolve an eye. But they had not been acting on the other animals to gradually be adapted to vision, in readiness for this eye. An animal will always be releasing an image into its sunlit environment, and the race to produce adapted images began. All those adaptations to vision that exist today were quickly conceived. The worm-like forms had to display armoured parts, warning colours, camouflage shapes and colours, or signs of the ability to swim so as to outmanoeuvre a pursuing enemy. Or, on the other hand, they could opt out of the visual environment and evolve bodies capable of burying themselves into rock crevices or other substrates. But after the initial chaos, further adaptations would become gradual – evolution would have settled down to its habitual pace.

That first eyed individual literally saw a whole new suite of niches open up. It observed areas of the sea floor in light and shade, which had previously been combined. But importantly it could easily identify the other animals sharing its environment. It could determine how far away they were, where they were heading, and how fast they were moving. At this point, nonetheless, there were to be few immediate consequences apart from a competitive edge this eyed individual had over other members of its species – it could find food and a mate more

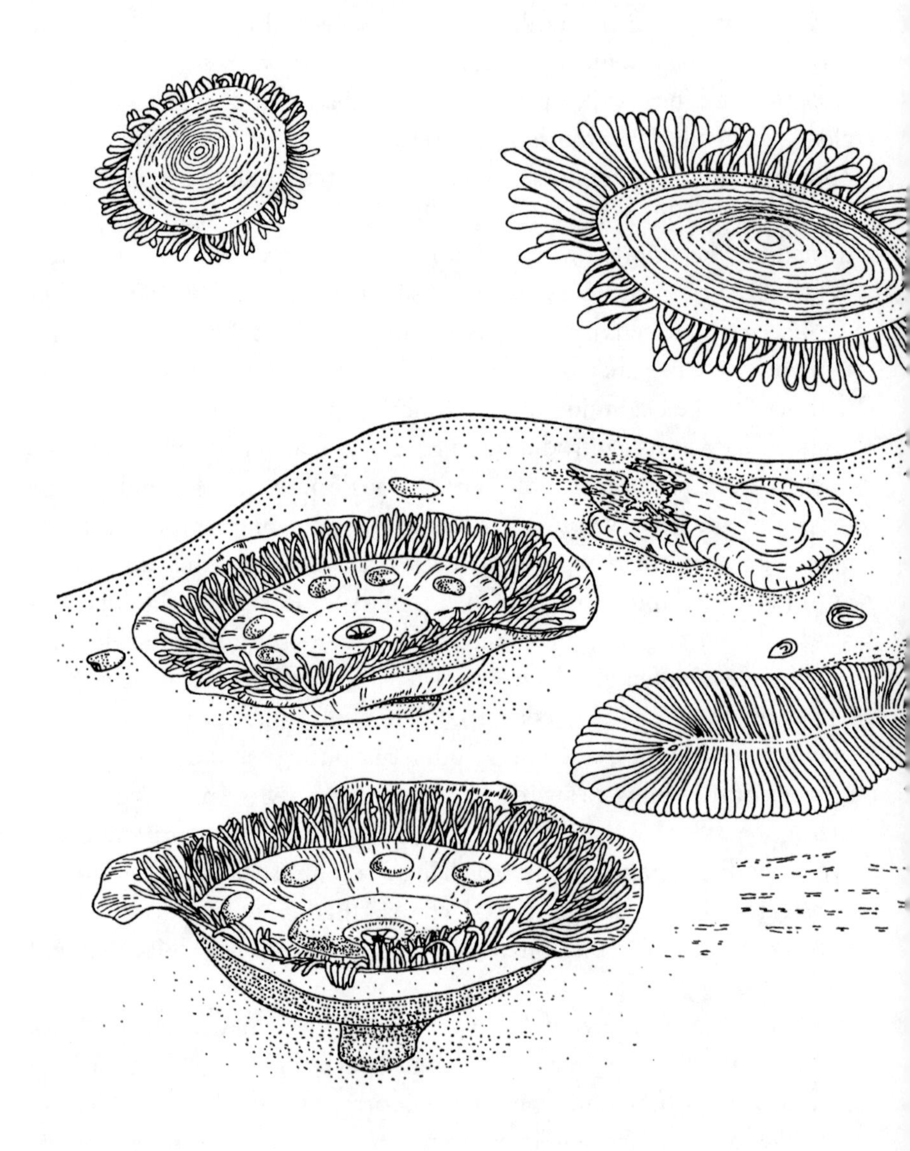

Figure 9.2 Soft-bodied multicelled animals living at the end of the Precambrian. This is how the most sophisticated light receptors of the time – eyes – would have pictured the Very Late Precambrian or Early Cambrian world, around 543 million years ago.

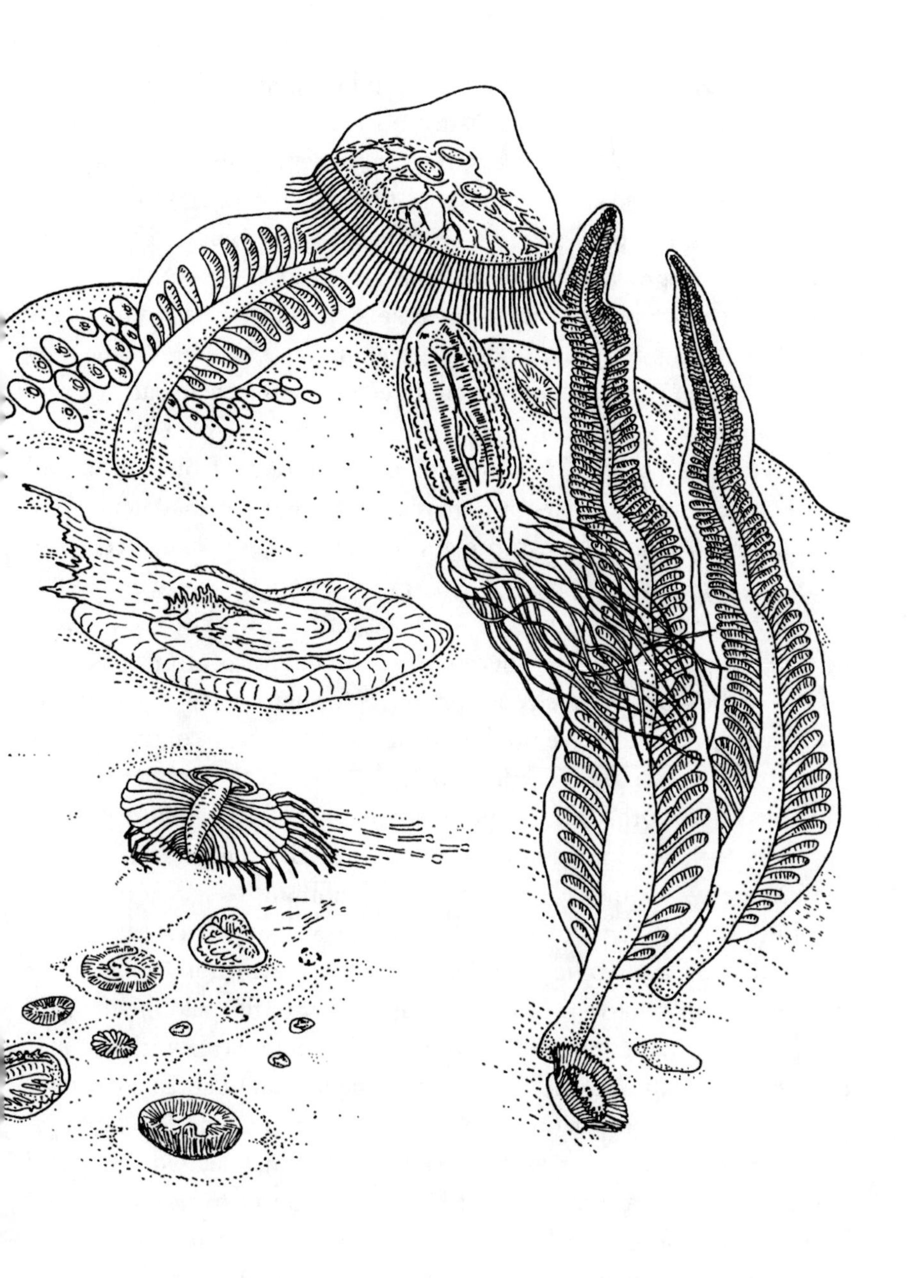

easily. This advantage would translate into retention in the species of those new genes that code for an eye. And soon all individuals of that proto-trilobite species would possess an eye, possibly making them a new species. But selective pressures for all multicelled life on Earth would have changed the moment that first eye opened, and the consequences of these would soon be realised. The next selective pressures were for active predation and its countermeasures.

The first eyed proto-trilobites must have been frustrated individuals. They had a taste for meat and were feeding on whatever scraps they came across on the sea floor, probably detecting the chemicals wafting from decaying 'food'. But now they could literally see a far greater potential. They saw their soft-bodied neighbours, from all animal phyla, as chunks of protein, or potential meals. But they had neither the mobility nor the jaws to capture and kill all of them. They needed to swim to capture those floating forms, and they needed stabbing mouthparts or limbs to perform their acts of murder. In other words, they needed hard parts. But considering the potential for proto-trilobites to take over the world, the selective pressures for hard parts were massive. And hard parts and active predation would follow, very quickly. Soon, proto-trilobites would become trilobites.

In seas across the globe trilobites with eyes and predatory limbs appeared at the beginning of the Cambrian. Active predation was born. Now there was a menace in the sea like nothing seen before. These trilobites set the scene for what was to follow, from *T. rex* carnivorising the Cretaceous, to lions in the Serengeti today. Another big factor in being a highly active predator was the ability of the trilobite to move up into open water – to swim. Today the bristle worms with the best eyes, the alciopids, are also the best swimmers of all the bristle worms – eyes are most useful if one is also highly mobile. The Precambrian predators in open water were those jellyfish which sensed the world mainly by touch. Animals cannot be adapted to touch, so this form of predation provided no selective pressures for the evolution of prey.

It really was the appearance of the trilobites that shook the world. Arrow worms were early Cambrian predators, but they were not known to be numerous and are rather tiny. In fact they hunted only small, planktonic prey and so could not have played a role in the

Cambrian enigma. And then, with some exceptions, no defences were evolved during the Cambrian explosion against the non-visual predators, such as the priapulid worms. The Cambrian explosion is really all about defences to visually oriented predation.

So when that first eye appeared, the potential for proto-trilobites to rule the world was recognised in the selective pressures acting on other animals. Selective pressures are invisible forces. No one is ever aware of them. One cannot 'urge on' evolution, even if one thinks one knows better. So as selective pressures for active predatory lifestyles mounted on the proto-trilobites, so did selective pressures for countermeasures build up on the other multicelled animals. And these pressures were massive too. Evolution is a balance, and the balance will not continue to tilt one way. With the exception of extinction, it continuously levels.

That first eye effectively created new niches for everyone, even though only the proto-trilobites could actually see them. Today, fishes do not know that they are silver to avoid predators. They evolved silver colouration to fill an available niche, one where large animals could live in mid-water if they were not visible to predators. And selective pressures targeted that available niche. So all those new potential niches at the beginning of the Cambrian, those areas of light and shade, were there for the exploitation of all. The rules were simple, but new.

Soon the free for all for trilobites was over. There was a new selection pressure acting on them – to avoid becoming prey. As they jetted through the water, and sprinted or skimmed over the sea floor, they came into contact with other trilobites. These other trilobites would have appeared as tasty morsels themselves. It became dog eat dog, or, rather, trilobite eat trilobite. The emphasis of trilobite evolution was shifting from eat to avoid being eaten. Some small Cambrian trilobite fossils have been found inside empty worm tubes. They were probably keeping out of sight of the larger trilobite hunters. But another evolutionary response was that the hard exoskeleton of the trilobites, that had granted their ability to swim, became endowed with armaments. And now, for the first time on Earth, armaments were ornaments. Let us return to the trilobite's soft-bodied food that was drying up, and in particular the reason why it was doing so. It was not simply that the trilobites were overconsuming.

The soft-bodied forms exposed on the sea floor started to become scarce because they were evolving. Before now they had been exposed to inactive predation only. This was a fairly inefficient process, in which maybe one in ten individuals would meet a sticky end. This may have been the sticky end of a predatory priapulid worm, or the sticky end of an anemone's tentacle, but a species can live with odds of one in ten. The remaining 90 per cent of individuals would have been safe – safe to carry the species into the next season. Stay out of the way of a priapulid or anemone and you are safe. A trilobite, however, will come looking for you. Things changed at the Cambrian border.

The most obvious requirement for adaptation to this new world of light would seem to be the possession of hard parts. This was precisely where evolution's emphasis was placed. Hard parts evolved for armour just as they had evolved in proto-trilobites to provide strong jaws. In most cases of ground-dwelling animals, their armour was directed towards attacks from above. This provides further evidence that active predators were swimmers – as suggested in Chapter 8, trilobites were probably the fishes of Cambrian seas. And then eyes themselves took off in the arthropod phylum not only to enhance a predatory lifestyle in their owners, but also to prevent them being eaten.

On close inspection of the fossil record, it becomes clear that it was the arthropod phylum that diversified most, or evolved the greatest range of hard parts, in the Cambrian. They were *the* active predators of the Cambrian, and eyes go a considerable way towards helping an animal become an active predator. The other thirty-three phyla that were to take on hard parts formed smaller armies. With the exception of molluscs and lamp shells, these other phyla were represented by relatively few species – species that saw their phyla through the Cambrian transitional period. This was achieved via adaptations to active predators with eyes, including the abilities to swim, hide in rock crevices, burrow efficiently or be protected by armour. Many of these adaptations required hard parts. Camouflage was probably another major adaptation but we have no evidence either for or against this – the Cambrian explosion was probably an event involving hard parts/shapes *and* colour. The other thirty-three phyla did not, however, evolve eyes in the Cambrian (with the possible exception of *Insolicorypha*, a

Cambrian bristle worm comparable to the swimming, eyed alciopids today). Maybe this could explain their reduced diversification in comparison with the eyed arthropods of the Cambrian. Eyes did evolve in five other phyla, becoming common only in the chordates and molluscs, but they evolved after the Cambrian – these five phyla remained eyeless during the Cambrian. For instance, according to the fossil record and evolutionary analyses, the group of eyed animals to which squid and cuttlefish belong did not evolve within the mollusc phylum until well after the Cambrian.

So it seems the evolution of hard parts everywhere, and ultimately the evolution of body forms of multicelled animals, was driven by active predators. This process *was* the Cambrian explosion. But it was triggered by the evolution of the eye. We are looking for that trigger rather than a detailed explanation of the event itself. The McMenamins' updated concept of food webs developing in the Cambrian is actually a description of the Cambrian explosion itself – but the event, not the trigger. The Cambrian explosion saw the writing of The Laws of Life as it exists today. The introduction of the first eye effectively tore up the previous Laws and gave rise to chaos among animals, creating a scenario without laws. It would have put evolution into top gear, perhaps moving it up from its lowest; fresh rules were required now. All animals needed to evolve to be adapted to vision before they were eaten, or before they were outwitted by their prey. The Early Cambrian thus became a race for adaptation to vision. This scramble for the newly available niches, this chaos during the writing of today's Laws of Life, *was* the Cambrian explosion. So finally we can be sure we have our answer. *The Cambrian explosion was triggered by the sudden evolution of vision.*

Life as we know it

The maximum depth of the Burgess Shale environment was 70 metres, and that is a sunlit environment where colour and ornaments are in operation today. *Hallucigenia* was a Burgess velvet worm and *Wiwaxia* a form of Burgess bristle worm, and both bore large, fearsome spines.

These animals lived on the sea floor and their spines were directed upwards into the water. They were aimed at predators swimming above them – they were armaments *and* ornaments (it is generally thought that bristle worms evolved spines in response to predators . . . predators with eyes, that is). The velvet worm *Asheaia* did not evolve spines, but rather formed an association with a sponge. Like comparable animals today, it probably evolved precisely the same colour as the sponge, or even stole the sponge's pigment. It is not easy to pick out a bristle star or crustacean living on a sponge today since their camouflage is perfect. Today's adaptations to visual predators probably evolved quickly during the Cambrian. From the beginning of the Cambrian up to today, the world has been adapted to predators with vision. For that same basic concept to have remained in place, unaltered, over some 540 million years illustrates just how powerful it really is. Powerful and stabilising.

The iridescence of *Wiwaxia*, *Canadia* and *Marrella* from the Burgess Shale probably functioned to deter predators. Like *Hallucigenia* and *Asheaia*, these animals could be considered slow-moving chunks of protein, but their protective spines would have appeared as iridescent, changing colours as predators approached. Changing colours or flashing lights are more conspicuous than a steady light, and so the visual warning created by the shape of the spines would have been enhanced. As explained earlier, both the shape and colour of the spines are adaptations to the vision of their predators – they are ornaments.

Using only their eyes, South American side-necked turtles can quickly estimate the nutritional value of potential prey in their environment. They will, however, only attack those animals that are vulnerable, regardless of their nutritional content. So a turtle will ignore a highly nutritional animal that is difficult to capture in favour of slow, easy pickings. *Wiwaxia* would have reflected all the wavelengths present in the Burgess Shale environment, and at least some of those wavelengths would have been used by its predators to achieve vision, so the iridescence would have been seen. It signalled that *Wiwaxia* was no easy picking. Basically, light display and vision would have been as important in Cambrian environments as they are in environments at comparable depths today.

I have spoken of the Cambrian as a transitional period, but it seems that the turmoil really was confined to those five million years or less at the beginning of the Cambrian – the Cambrian explosion. After this, after everything had adapted to vision, stability began to replace the chaos. The arthropods *Isoxys* and *Waptia* are genera well known from Burgess Shale fossils, 515 million years old. But they were discovered also in the Chengjiang site of China, 525 million years old. So those genera existed for a considerable period of time – at least ten million years. *Anomalocaris* is known from an even longer stretch of the Cambrian, and the list continues, with life appearing quite well conserved following the Cambrian explosion. Vision entered the Earth with a bang, or a flick of the light switch, and then things settled down. The introduction of vision triggered a scramble to occupy the new niches that had opened up as a consequence. Once all those niches had been filled, micro-evolution resumed. Think of the dinosaurs. Dinosaurs occupied the top predatory niches and kept mammals low down in the food pyramid. Mammals could only take over once the dinosaurs were removed – when niches are filled there is stability in the system, a stability which resists change.

Why vision and not other senses?

Light is only one of a suite of stimuli in today's environments. Maybe the other stimuli should be considered in relation to the Cambrian explosion in much the same way as light has been considered in this book. Could the light switch really be the sensory switch, where all senses were introduced on Earth at the dawn of the Cambrian explosion? Or did the other major senses of animals today evolve gradually before and after the Cambrian explosion – were they subject to micro-evolution rather than macro-evolution?

First of all, what are the senses other than vision? A sense is the ability to detect and be conscious of the outside world. A sense involves a stimulus and a detector. Excluding eyes and vision now, detectors are usually one of two types – chemical or mechanical. Magnetic detectors also exist, which track the direction of the Earth's magnetic field. The magnetic sense is best known in insects and chordates, such as the homing pigeon, which can determine its geographical position by using a 'magnetic map'. Some fishes, notably sharks, use the magnetic sense to hunt prey, but usually this sense is employed for orientation purposes. It is not known to have played a role in predator–prey relationships until after the Cambrian, and so can be excluded as a possible cause of the Cambrian explosion.

A case of micro-evolution

Chemical detectors detect chemicals and give rise to the senses of taste or smell. They contain nerve fibres that produce electrical impulses when contacted by specific chemicals. In its most rudimentary form, a nerve fibre terminates at the outer surface of the host animal and is free to be stimulated by chemicals that contact the animal. This can become more complex in different ways. Increase the density of the exposed nerve fibres and the animal becomes more sensitive to the chemical detected. Mix these nerves with different nerves sensitive to other chemicals and the animal becomes receptive to an array of chemicals. Then hold the nerve fibres away from the surface of the animal, so they protrude into the environment, and the animal becomes increasingly sensitive to chemicals. This is because the sticky boundary layer that surrounds animals acts less as a barrier to chemicals. Also, at a distance from the body surface there is less 'noise' or background chemicals coming from the host animal itself. Nerves that enter the environment can be protected within hairs. The detectors can range from a single nerve fibre within a single-pore hair to bundles of nerve fibres inside a hair with many perforations. And increasing the density of such hairs on the animal can increase the complexity and sensitivity of the hair system further. But the same pattern emerges however sensitivity is increased. This is a pattern that prevents chemical detectors from causing an explosion in evolution.

For every complex, highly sensitive detector of chemicals there is a well-defined and gradual evolutionary path. A detector in the form of a clump of hairs would have evolved from a few hairs, which in turn would have evolved from a single hair. That single hair would have evolved from a single bump in the outer surface of the animal, which would have derived from a nerve penetrating the flat surface of a previous ancestor. But importantly, this path is rather smooth. In other words, the evolution of smell and taste over geological time was linear – it involved a series of numerous but gradual transitional stages. The evolutionary ride was also smooth.

Perhaps 'linear' and 'smooth' are exaggerations. Hairs, for instance, involve hard parts, and, as we know, hard parts did suddenly emerge overnight on the geological timescale. So at this point in history, the sensitivity of smell and taste detection may have jumped up a notch. But that jump would not have been monumental because the information supplied by the nerve fibres did not suddenly increase *several* times. And this particular jump was probably the biggest of all for chemical detectors, but it happened *during* the Cambrian explosion, and so could not have been the trigger. The evolutionary ride for smell and taste, nonetheless, would have been littered with modest bumps.

Mechanical receptors are so called because they sense physical movement in the environment. They contain nerve fibres that produce electrical impulses when they are themselves moved. This happens as a consequence of contacting an object or movement in the surrounding water or air. Mechanical receptors are responsible for the senses of touch, hearing/vibration detection, and gravity, temperature and pressure sensitivity. Like chemical detectors, mechanical receptors occur in a variety of shapes and sizes, but they also demonstrate that same evolutionary pattern. In theory, the evolution of mechanical receptors took place in small steps.

The evolution through geological time of chemical and mechanical receptors cannot be compared with the evolution of light detection. There is no event in the evolution of receptors of other senses that can match, or even come close to, the evolution of the lens. Chemical and mechanical detectors certainly would have become more efficient throughout the Cambrian explosion, but not to the extent that they would have changed the entire behavioural system of animals. There is no case of a receptor suddenly changing in efficiency 'a hundredfold', like the change from a light-sensitive patch to an eye capable of producing visual images. Here lies the fundamental difference between light detectors and the receptors of other stimuli – those of other stimuli still work at their intermediate stages of complexity and efficiency. The evolution of receptors for stimuli other than vision can theoretically show a linear progression, but a light perceiver with an inadequate lens has little advantage over one with no lens. The theoretical intermediate stages of a lens increase light perception only slightly, but when a complete, fully focusing lens is formed, the increase suddenly becomes vast. This leap in efficiency is so glaring that it led Darwin to single out the eye as the thorn in the side of evolutionary theory. But it also indicates that if the evolution of one type of receptor could trigger the Cambrian explosion, that receptor would be an eye. The evolution of the eye from a rudimentary light perceptor – that single jump from 'not to see' to 'see' as described in Chapter 7– was a small step for anatomy but a huge step for animal behaviour.

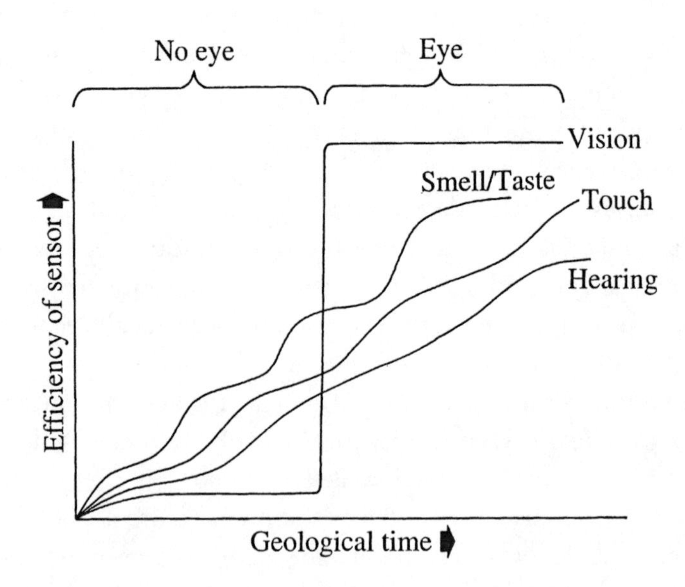

Figure 9.3 Graph showing the very approximate evolution of receptors for different stimuli throughout geological time. Vision is the only sense that can divide geological time into two distinct phases.

A presumed Precambrian history of electrical wiring and information processing

There is the evidence to be gained from the history of the nervous system and brain capacity to process the information collected by the detector. We now know that the eye *did* suddenly appear at the beginning of the Cambrian, and that it instantly became a common characteristic. A considerable number of Cambrian arthropods possessed eyes, so they must have worked then as they work today. For an eye to work, sizeable brain and nerve cables are required, and these were in part borrowed from other senses. This is the most conceivable way in which an eye can suddenly achieve vision, after its leap from simple progenitors, the light perceivers. What does this borrowing tell us? It indicates that at least some senses had evolved to a reasonable degree of sophistication before the Cambrian, so that they had established a nerve network including brain space. In turn this means they could not have triggered the Cambrian explosion, which leads us on to another interesting detail in the evolution of senses other than vision, a detail that emerges when the different animal phyla are compared.

Evidence from evolutionary history

Up to now we have been interested in the evolution of senses throughout geological time, but we should forget about time for a moment and look again at the evolutionary tree of multicelled animals, the tree that shows the order of genetic mutations and establishment of internal body plans. This will reveal the evolution of sensory systems throughout the evolution of phyla, rather than time. And generally we find that the first phyla to branch off the tree are represented by species today with simple nervous systems.

Mechanical reception, specifically touch, occurs in sponges by general cellular irritability. In the more evolutionary derived phylum, the flatworms, it occurs by the stimulation of free nerve endings. Even more derived phyla contain more sensitive mechanical receptors, which can detect pressure waves created by the low-frequency vibrations of distant objects. The most derived phyla can even detect sound waves or high-frequency vibrations by means of special phonoreceptors.

Other senses show a similar pattern of gradual development throughout evolution. Sponges are without specialised chemical receptors, but again taste/smell is a general property of the body surface. Sponges respond to chemical irritants by contraction of the opening to the body cavity, which restricts the flow of water and the irritant through the sponge. But it is the contractile cells surrounding the opening that respond directly to the irritants – no specialised receptors are involved. The next phylum to branch off the evolutionary tree includes sea anemones and jellyfish. These animals possess simple chemical receptors in their mouths and tentacles, which can distinguish between different food types – anemones and jellyfish prefer some foods to others. The chemical senses are even better developed in molluscs, echinoderms (starfish phylum), arthropods and other more highly derived phyla, where they are also used to locate food sources.

The capacity to discriminate temperature differences is slight in members of most phyla, but thermal sensitivity is acute in most chordates, one of the more highly derived phyla. Although within the chordates fishes are the best-known marine animals in terms of sensitivity to pressure changes, most swimming members of other phyla are also known to respond to pressure variations. This includes some of the least derived phyla – jellyfish and comb jellies – as well as bristle worms and arthropods. A reaction to gravity is also shared by most phyla. Comb jellies and some jellyfish are equipped with this sense, along with bristle worms, echinoderms (including sea cucumbers), lamp shells and arthropods.

Sound waves travel readily through water, with greater speed and less dampening than in air. But hearing as such is an adaptation of the chordates. Members of other phyla may detect sound waves, or at least vibrations of some description in the water, by means of less complex organs. The reflex of some bristle worms to underwater sounds is to start, and certain crabs are known to produce sounds. The means of sound detection, which is certainly very limited, is unknown in these animals, but large, specialised acoustic organs are absent. In the case of crabs, gravity detectors may be involved. Insects began their history as deaf animals, so the noisy cicadas, crickets and grasshoppers had to evolve sound detectors in addition to sounds. This dual evolution is a lengthy process, requiring fine-tuning of both receptor and transmitter. Importantly, this sense has little direct effect on its neighbours from other animal phyla. And it was the effect of vision on *all* phyla, not just those with the detectors, that made eyes instrumental to the cause of the Cambrian explosion.

So at the phylum level there is a definite relationship between sensory efficiency and the branching point from the evolutionary tree, light detection excluded. Sponges, the least derived phylum, possess simple forms of mechanical and chemical receptors. The next phyla to branch from the tree, the cnidarians (including jellyfish) and comb jellies, again have simple forms of touch receptors and slightly more sensitive chemical receptors, but also reasonable pressure and gravity receptors. Flatworms, one of the next most derived phyla, possess further improved mechanical receptors. But the more highly derived phyla show a general improvement of most types of sensory receptors. This is to be expected. The trend is one of increasing sensory perception with increasing complexity of the body, and this includes brains and nervous systems, those attributes vital to sensory perception. Again, this suggests that the senses other than vision evolved gradually, beginning their history before the Cambrian. Eyes, it would seem, are the oddballs of sensory evolution.

An unavoidable presence

Finally, there is the argument, touched on several times already, that the sense of light detection is different from other senses because of its stimulus. In most environments, sunlight is present, and any animal will leave its optical signature, or image, in that environment. This image is ripe for detection. So to adapt to vision, an animal *must* evolve a response in terms of adapting its visual appearance, whether it is warning shapes and colours, camouflage or hiding behind physical barriers.

Most common senses other than vision begin with a stimulus created by an animal. So if an animal does not create the stimulus, it can't be detected. And then chemical receptors and, to some extent, mechanical receptors, are often finely tuned to detect only a narrow range of the potential stimuli. So animals can evolve to avoid only that specific range. It is not so simple to adapt to vision, however, since eyes usually detect most of the stimulus range, or spectrum, in their environment. I saw this principle in action, curiously enough, while writing this chapter, when I observed a jumping spider take on a 'flesh fly' twice its size. The spider was positioned on a wall, against which it was well camouflaged. The fly landed just 10 centimetres from the spider but did not detect its presence. The fly has excellent chemical receptors, but not specifically for the smell of jumping spider. And because the spider was neutral to vision, the fly could not sense it. As the spider approached the fly, however, it was compelled to make movements. These movements translated to changes in its visual appearance and were detected by the fly, which flew off. Fortunately for the fly, the sun always shines, and the spider cannot help but leave a signature in the visible spectrum. Even evolution cannot provide a perfect solution to that problem.

In essence, *all* animals must adapt to light, but this is not the case for other stimuli. And to adapt to a radical advance in chemical perception, for instance, an animal must reduce the chemicals it exudes to a minimum. But this change would have little to do with hard external parts. In fact most changes of this nature would occur inside an animal, in its chemical processes. So a revolution in chemical reception could not have caused the Cambrian explosion – the evolution of external parts.

Eyes bring new opportunities

From another perspective, adaptations to vision do affect other senses. As the door is closed to visually oriented predators, it is opened to predators mainly employing other senses. Hard, protective shells are often ornaments to predators with eyes, and signal that an attack would be a waste of energy and might even harm the attacker. But blind predators are oblivious to this signal. The shelled animals have evolved best to counterattack by the greatest threat in the water – highly active predators with eyes. And in doing so they created a new niche – one for less active predators. Enter starfish, creatures that are blind but can prey on less mobile but even well-protected animals. Starfish rely on smell and touch to locate their prey, which they then smother until an opening to the soft, edible parts is located. But this is only possible because animals can't be adapted to everything, and they are generally adapted to counter the greatest threat. Other threats, then, can enter the system through the back door. This back door, however, was once the front door.

Near-final thoughts

It should be remembered that there was never really a race waiting to begin in the Precambrian, a race to attain eyes. That's not the way evolution works, and would represent a teleological view. Rather, something happened in the environment one day that changed the rules. Then selective pressures changed either in their direction or size. Evolution works by adaptive radiation, usually caused by a change of some description in the environment. In *The Theory of Evolution*, John Maynard Smith explained further that 'when a reversal or change in the direction of evolution has occurred . . . it perhaps more often [reflects] a change in the methods of exploiting that environment'. Whichever way you look at it, the appearance of eyes was the biggest change in the environment of all, even for those blind animals. But although vision can be found in only six of the thirty-eight phyla today, over 95 per cent of *all* animal species, taking account of *all* phyla, have eyes. Eyes certainly proved a significant method of exploiting an environment.

In his 1992 review on 'The Evolution of Eyes', Michael Land began with the statement 'Since the Earth formed more than five billion years ago, sunlight has been the most potent selective force to control the evolution of living organisms.' This is true for life in general, particularly those forms that photosynthesise, but for *animals*, barring the inefficient sense of simple light perception, it is true for the past 543 million years only. Although the figure of 'five billion years' does not apply to animals, Land's statement otherwise supports my inferences made in Chapters 3 to 5. But of greater importance to this book is the understanding of why 'five billion' does not apply to animals. If one divides the history of the Earth into pre- and post-eyes, then considering the power of vision – generally the most potent selective force for animals today – its day of birth must have been a monumental event in the history of life. Forgetting the Cambrian explosion for a moment, the evolution of vision, that opening of the first eyes, must have caused a remarkable change in the way life works, particularly with respect to external forms of animals. That this day coincided with the day animal life began to explode seems more than a coincidence.

In his conclusion to *Origin*, Darwin wrote:

> It is interesting to contemplate a tangled bank, clothed with many plants of many kinds, with birds singing on the bushes, with various insects flitting about, and with worms crawling through the damp earth, and to reflect that these elaborately constructed forms, so different from each other, and dependent upon each other in so complex a manner, have all been produced by laws acting around us.

Walking around the extensive garden at Down House, I noticed a similar diversity. But I should have seen more. According to a book about local fauna, there is much more to see in the countryside visible from Darwin's garden paths. Set against the white background of the pages of a book, the rabbits, several species of common birds, further species of even commoner beetles, frogs, snakes . . . many local animals would seem easy to spot. But against their natural backgrounds, they simply cannot be seen. They are adapted to the light in their environment – they maintain a low visual profile. Even though the birds could be heard, they could not be seen. One sees mainly plants – and plants generally abstain from adapting their colours to avoid the attention of animals.

If Darwin could have travelled back in time, donned Scuba gear and walked through the Late Precambrian seas, he would have seen animals from all phyla everywhere. He would have noticed worms and other soft-bodied forms, including those ancestors of the mammals, crawling and floating in front of his very eyes. Simply, in the Precambrian, animals were not adapted to vision, and there was no danger in being incidentally bold. That could not happen today.

10

End of Story?

> The eye of the trilobite tells us that the sun shone on the old beach where he lived; for there is nothing in nature without a purpose, and when so complicated an organ was made to receive the light, there must have been light to enter it
>
> JEAN LOUIS RODOLPHE AGASSIZ, 'Geological Sketches' (1870)

So the evolution of vision via that very first eye in a trilobite triggered the Cambrian explosion. This is the answer to the problem – the Cambrian enigma – I set out to solve. In 2000, I presented this solution at a Royal Institution Lecture in London, where it sparked many questions. I could answer all of them . . . except one. The Light Switch theory also succeeds in posing a further question. As one door closes, it seems that another one is opened.

At the end of my Royal Institution lecture came the question, 'What triggered the evolution of the eye?' I believe this does require an answer, that we should not assume an eye was always going to evolve as soon as the genetics and building materials in an animal became appropriate (a teleological view). Recently this question has attracted the attention of geologists and meteorologists, who have begun to search for an answer. Logic suggests the solution must lie in an event which led to an increase in light levels at the Earth's surface just prior to the Cambrian. This would suddenly enhance the selective pressures for an eye to evolve. But what was that fateful event, which indirectly changed the course of the history of life on Earth?

The first eye must have evolved in response to an increase in *sunlight*, a factor independent of evolution – bioluminescence (light generated by animals) would not have evolved significantly until there was an eye to see it. And indeed the geologists have revealed an increase in sunlight levels at the Earth's surface precisely at the very end of the Precambrian. Due to its direct relationship with the Earth's magnetic field, an increase in luminosity is proportional to an increase in the elements carbon-14 and berylium-10 preserved in the rocks. And temperatures increased on Earth at that time too. So we have our answer, or at least part of it – eyes evolved when the dominant selection pressure for an eye stepped up a gear. But we still seek a factor that caused an increase in sunlight levels. Light passes from the sun, through the space of our solar system (the interplanetary medium), through the Earth's atmosphere and through the sea (remember, Cambrian life was exclusively marine). So for sunlight levels to increase at the Earth's surface, one of two events must have taken place: either the sun's light output increased, or the media between the sun and Earth's sea floor became increasingly transparent.

Through theories of stellar construction, it has been well established that the sun was between 25 and 30 per cent less luminous 4,600 million years ago than it is today. But the pattern of this increase in light output is unknown, although it is assumed to have been gradual. Because of the immense time period under consideration, a gradual increase, or even a stepwise increase, translates to a very minor boost in sunlight during the few million years prior to the Cambrian explosion. But it is still possible that sunlight levels rose to a critical level at the end of the Precambrian – critical in that light sparked new reactions within the Earth's atmosphere that led to increased transparency. And this brings us to the second possibility for a rise in Earth's measure of sunlight.

Certainly, the content of the Earth's atmosphere affects its transparency to light – different elements absorb sunlight to different degrees. And the atmospheric contents *have* changed throughout geological history. Some meteorologists suggest that a blanket fog (with various possible sources, including volcanic activity) cloaked the Earth's surface in the Precambrian, thus blocking out a high proportion of sunlight like a giant umbrella. So the lifting of this fog at the very end of the Precambrian would have greatly increased light levels at the

Earth's surface. Precisely how the fog lifted is another issue altogether. One suggestion is again linked to a slight but critical increase in radiation from the sun. The merest increase in solar output and the blanket fog becomes transparent water vapour. So, almost overnight in geological terms, the Earth has clear skies and a line of sight. This would seem the tidiest explanation for a sudden increase in sunlight at the end of the Precambrian. But there are other possibilities.

So far I have considered changes in transparency *within* the Earth's atmosphere. But are there extraterrestrial possibilities? Could there have been an event that reduced sunlight absorption between the sun and the Earth? There may have been, and its origins could exist deep within our galaxy.

Earth lies within a solar system that lies within a galaxy. The stars in our galaxy are clustered to form the shape of a 'plate' with a bulbous centre. But this galactic plate is not even – outside the central zone there are four 'arms' that spiral (logarithmically) out towards the edges. Although it has always existed near the edges of the plate, our star – the sun – has not always occupied the same position within the galaxy. It has moved around through time, passing in and out of the spiral arms. It streams through the arms at a speed of 68 kilometres per second, and spends tens of millions of years within each arm during crossover. And to a lesser extent it also moves up and down within an arm – the plate that is our galaxy does have some thickness.

As our solar system moves into a spiral arm, it encounters large, concentrated complexes of molecular gases and dust, but also a greater density of stars – it moves closer to other stars. Sometimes stars explode, causing 'supernovae', and at some stages in its history the Earth has been relatively close to supernovae. Supernovae probably represent the most violent events in our solar neighbourhood during geological history. And, of relevance to our discussion, they cause changes in the interplanetary medium of our solar system.

Supernovae cause the absorption of visible light by the formation of nitrogen dioxide. So in turn they reduce the light levels at the Earth's surface. Additionally, while passing through a spiral arm, our solar system could also traverse a dense 'Oort cloud' that would raise the sun's brightness but also make the Earth's atmosphere more opaque.

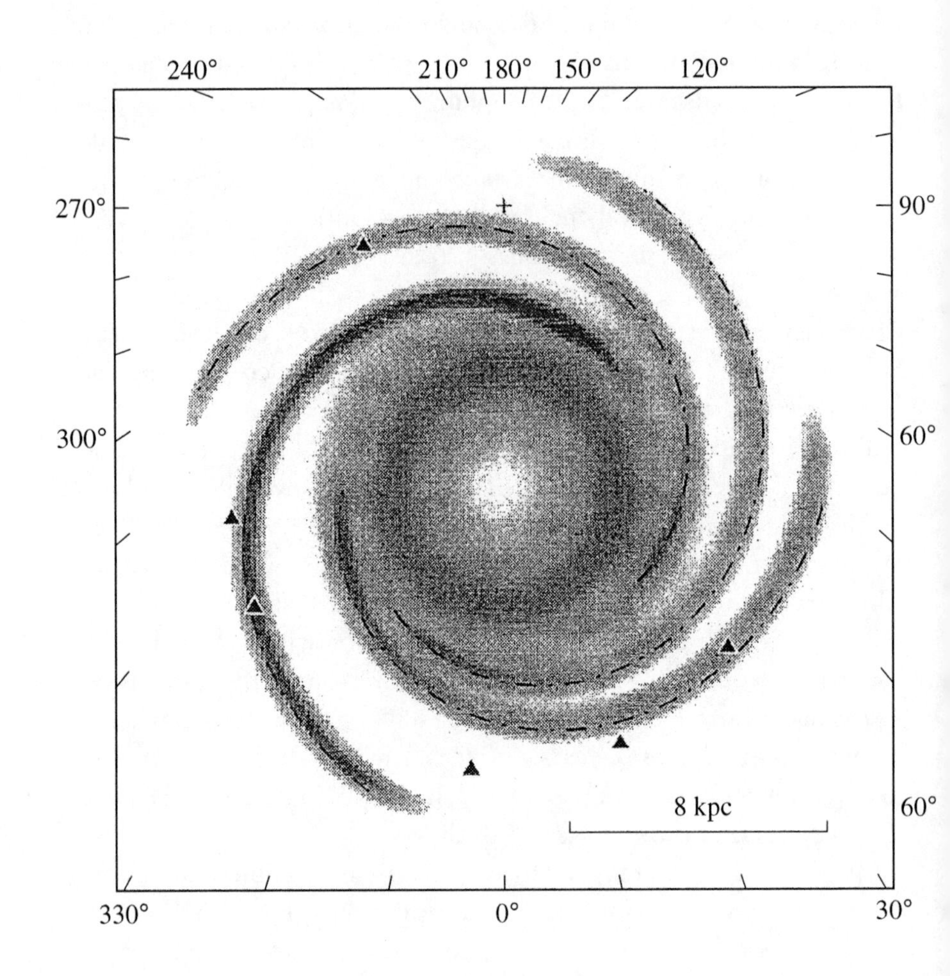

Figure 10.1 Face-on view of our galaxy. Counterclockwise from the Sun (cross at top) are the Sagittarius–Carina arm, Scutum–Crux arm, Norma arm and Perseus arm. Triangles mark the times of the major post-Cambrian extinctions (modified from a paper by Erik Leitch and Gautam Vasisht). Some researchers believe the movement of our solar system into the spiral arms had an effect on these extinctions (such as a consequential encounter with giant meteors). The effect of unwinding is indicated by the dot-dashed lines defining the centroids of the arms for an unwinding of 1°, 4° and 8° for the first three arms, respectively.

Again, the net effect would be a reduction in light levels at the Earth's surface. So as our solar system departed from a supernova or an Oort cloud, the Earth would have become a brighter place. Maybe this increase in sunlight could have been the enhanced selection pressure for eye evolution. This situation is comparable to, or even the same as, the 'blanket fog' scenario discussed earlier.

Supernovae can also cause ozone depletion in the Earth's atmosphere via enhanced ionising radiation and cosmic rays. This, as we well know from the hole in the ozone layer today, increases the portion of some ultraviolet wavelengths reaching the Earth's surface. But these ultraviolet wavelengths are not the same as those employed in vision – they are shorter, and are a concern for their damage to animal tissues rather than visual ammunition. And in terms of directly increasing the sunlight reaching Earth's surface, a supernova emits only a flash of light – nothing long-lasting enough to be a selection pressure for evolution. So its effect on evolution could be only via changes in the interplanetary medium or within the Earth's atmosphere. But maybe this was enough to give evolution a nudge in a particular direction. The next stage of research to be conducted in this area involves timing; did the Cambrian explosion coincide with the Earth's passage through the spiral arm of the galaxy? That remains to be discovered.

Finally, we should consider changes in sea transparency. In terms of quality of light, or colours, today the sea acts as a narrow filter. Only a restricted range of wavelengths – mainly in the blue region – pierce seawater well, and the rest are absorbed or scattered. But change the mineral content of the sea and this filter may move within the spectrum or even widen. Could there have been an event at the Earth's surface that released minerals previously locked in rocks? Today the lakes in the Canadian Rockies are a stunning emerald green. Glaciers have stirred up the rocks in their paths and so changed the mineral content of the waters encountered over time, and consequently shifted the light wavelengths reaching the lake floors. So the waters at the edges of the oceans, the hosts of the Cambrian explosion, could potentially have changed in mineral content and light transparency too. Maybe, at the end of the Precambrian, the light in shallow seas suddenly included ultraviolet light – the ultraviolet wavelengths employed in vision today.

That would be interesting because it could have complimented the very first eye.

We are beginning to learn more about those private ultraviolet wavelengths used by some animals excluding ourselves. We cannot see ultraviolet light because our lens absorbs it. Earlier in this book I described how we became familiar with nature's ultraviolet patterns – they were captured on camera film. Although an ordinary glass camera lens absorbs ultraviolet light, a quartz lens is extremely transparent, particularly to those ultraviolet wavelengths used for vision by arthropods and some other animals today. Quartz also formed the lenses of trilobite eyes. So that first eye could potentially see ultraviolet light, providing it possessed ultraviolet sensitive cells in its retina. And that was likely, since retinal cells for blue light also detect some ultraviolet in animals, including ourselves (people with artificial lenses can indeed see in the ultraviolet). Because blue light would have been optimal in the Cambrian seas, trilobites would certainly have possessed blue-sensitive retinal cells.

Although the sea is not particularly transparent to ultraviolet light today, there are some shrimps and other animals which have the ability to see these wavelengths. In fact this finding is becoming increasingly common. An increase in ultraviolet transparency in seas at the end of the Precambrian could have been due, again, to a change in mineral content, but also to a reduction in 'particles' that scatter light. These particles scatter shorter wavelengths of light, representing blues and ultraviolet, much more than longer wavelengths, representing the red end of the spectrum. So without these particles, the waters below the very surface of the sea would have contained more ultraviolet wavelengths available for vision.

Similarly, atmospheric events could have caused an increase in usable (for vision) ultraviolet *reaching* the sea. The 'particles' that scatter sunlight in the Earth's atmosphere cause the sky to appear blue – and ultraviolet. Meanwhile the remaining wavelengths pass directly through the scattering layers, and we see them during a sunset where they appear orange and red. So variations in the density of these scattering particles can shift the emphasis of the Earth's spectrum from red and orange to blue and ultraviolet. But because we don't know precisely

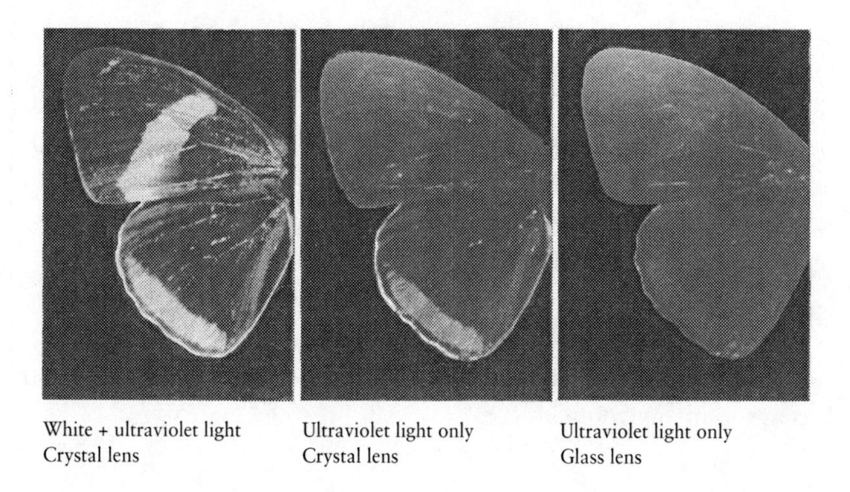

Figure 10.2 From left to right: a butterfly wing photographed in black and white through a crystal lens under white plus ultraviolet light; through a crystal lens under ultraviolet light only; and through a glass lens under ultraviolet light only. To the human eye each wing appears black with two blue stripes. These images reveal that the lower stripe also reflects ultraviolet light, which transmits through the crystal lens but is absorbed by the glass lens.

which colours the first eye saw, we must end our search for the wavelengths that changed to provide an enhanced selection pressure for vision.

Now we are left to consider only the general quantity, or brightness, of sunlight as a selection pressure for eye evolution. But again, a mineral change in the water is the most likely explanation for increased light transmission in general (an alternative could be the clearing of dense algal blooms). So we require still an event that could have led to this. Maybe it is time to re-open the evolutionary file for Snowball Earth.

In Chapter 1, I described how the Earth passed through spells where it was covered, or nearly covered, in ice a kilometre thick. Certainly the retraction of this ice could have stirred up minerals in rocks on a grand scale. As those huge ice sheets traversed the land, they would have ripped open the surface layers of rocks and absorbed minerals, transporting them to the sea. Unfortunately, though, the timing is a little out. The

Cambrian explosion took place between 543 and 538 million years ago, and the last Snowball Earth event ended 575 million years ago at the latest. So there is a difference of at least thirty-two million years between these two events. This might be just too great – theoretically an eye can evolve within half a million years. So I still believe that the last Snowball Earth event should be coupled to the Precambrian 'surge' in evolution rather than with the Cambrian explosion.

Research in this area of the geological history of media transparency is still in its infancy; hence my discussion of this subject has been brief. In the future it is to be hoped that all will become as clear as the Late Precambrian environment itself.

A final word

The Light Switch theory is a consequence of recent fossil finds and evolutionary analyses (although the philosophy of colour today weighs in heavily, too). There remains an imperfection in the geological record that is still to be reckoned with, but it no longer looms before us as it did in Darwin's days. Palaeontologists today are striving to fill the ever narrower gaps in the fossil record, searching all corners of the globe for new species that lived near the time of the Cambrian explosion.

Originally I was afraid that the Light Switch theory might appear far-fetched, particularly since most alternative theories had been heading in very different directions. Eyes the cause of the Cambrian explosion? How ridiculous! But it was the amalgamation of modern biology with Cambrian palaeontology that finally settled my nerves. Now, after considerable contemplation of the power of vision today, I am convinced that the evolution of that very first eye must have been a monumental event in the history of life on Earth. For this fact alone I am happy to share my ideas with a wider audience. Whether that introduction of the eye really did coincide with the beginning of the Cambrian explosion should be answered with greater precision as new fossil finds are unearthed from near that Early Cambrian border. But at this stage in our knowledge, this relationship appears remarkably close.

My final reassurance that the Light Switch theory is both a judicious and logical one came from the editor of a newspaper. James Woodford, a journalist with the Australian newspaper the *Sydney Morning Herald*, wrote a comprehensive article on my theory. This made the front-page headlines, and gave the newspaper's editor cause for concern. The night before publication, and just before the article and the paper went to press, James received a question from his boss. The question was, 'Are you sure this has not been said before?' That was extremely comforting. It meant that this was an obvious answer. In fact it was so obvious that it had little scientific merit – anyone could have come up with it. True. Now *I* think that this is the obvious answer.

Recently I went swimming off the coast of Sydney. Here I encountered a group of cuttlefish similar to those that had initially woken me up to biodiversity, as described in the first chapter. Again the cuttlefish surrounded me in an arc and displayed spectacular colour changes. Again they looked at me with their large sophisticated eyes, and flashed their sophisticated colour display, as if confirming the importance of light in nature. Yes, I thought, vision has really entered the behavioural system of animals. Then I noticed a crab on the sea floor. I zoomed in on its eyes, and reflected: *the origin of those arthropod eyes had a lot to answer for . . .*

Index

Page numbers in *italics* refer to illustrations and figures in the text.

Andrew Parker received his Ph.D. from Macquarie University in Sydney while working in marine biology for the Australian Museum. He became a Royal Society University Research Fellow at Oxford University's Department of Zoology in 1999, and is an Ernest Cook Research Fellow of Somerville College, Oxford, and a Research Associate of the Australian Museum and University of Sydney. He has published numerous scientific papers on topics as diverse as optics in nature, biomimetics and evolution. He lives in Oxfordshire, England.

CPSIA information can be obtained at www.ICGtesting.com
Printed in the USA
LVOW100001200112

264655LV00009B/27/A